SENSOR ARRAY

Edited by **Wuqiang Yang**

Sensor Array
Edited by Wuqiang Yang

Published by InTech
Janeza Trdine 9, 51000 Rijeka, Croatia

Publishing Process Manager Jana Sertic

Technical Editor Teodora Smiljanic

Cover Designer InTech Design Team

Additional hard copies can be obtained from orders@intechopen.com

Sensor Array, Edited by Wuqiang Yang
p. cm.
ISBN 978-953-51-0613-5

For EU product safety concerns:
IN TECH d.o.o., Prolaz Marije Krucifikse Kozulić 3, 51000 Rijeka, Croatia,
info@intechopen.com or visit our website at intechopen.com.

Meet the editor

Professor Wuqiang Yang (FIET, FInstMC, FIEEE, CEng) received all his degrees (BEng, MSc, PhD) from Tsinghua University. After 3 years as a Lecturer at Tsinghua University, he joined UMIST in 1991 and currently is a Professor at University of Manchester. His research is focused on electrical capacitance tomography. He has published 300 papers, and holds 10 patents. He is an editorial board member of 5 journals, a guest editor of several journal special issues, and reviews papers for 40 professional journals, including 6 IEEE Transactions. He has been invited by many universities, research institutions and international conferences worldwide to give lectures, seminars, or keynotes. He is a visiting professor at 6 universities. Since 2002, his biography has been in Who's Who in the World. He is currently an IEEE Instrumentation and Measurement Society Distinguished Lecturer.

Contents

Preface

It is my great pleasure to be an editor for InTech Open Access publisher, one of the most successful Open Access book publishers in the fields of science, technology and medicine.

Sensor arrays are used to overcome the limits of simple and/or individual conventional sensors. Obviously, it is more complicated to deal with some issues related to sensor arrays, e.g. signal processing, than the conventional sensors. Some of the issues are addressed in this book, with emphasis on signal processing, calibration and some advanced applications, e.g. how to place sensors as an array for accurate measurement, how to calibrate a sensor array by experiment, how to use a sensor array to track non-stationary targets efficiently and effectively, how to use an ultrasonic sensor array for shape recognition and position measurement, how to use sensor arrays to detect chemical agents, and applications of gas sensor arrays, including e-nose. This book should be useful for those who would like to learn about recent developments in sensor arrays, in particular for engineers, academics and postgraduate students studying instrumentation and measurement.

I would like to take this opportunity to thank all the authors, who made valuable contributions with the chapters in this book. Without their hard work, it would be impossible to publish this book. I would also like to thank Ms Jana Sertic for her efficient management.

Professor Wuqiang Yang
Professor of Electronic Instrumentation,
School of Electrical and Electronic Engineering,
University of Manchester, Manchester,
United Kingdom

1

Passive Source Localization of Sensor Arrays

Junli Liang and Ding Liu
Xi'an University of Technology, Xi'an,
China

1. Introduction

Passive source localization is a key issue in sensor array signal processing such as sonar, radar, wireless communication, microphone array speech processing, seismology, electronic surveillance and medical imaging, and thus receives significant attention. Although a variety of advanced algorithms, for example MUltiple SIgnal Classification (MUSIC), Estimation of Signal Parameters via Rotational Invariance Techniques (ESPRIT), and Propagator Method (PM), have been developed, there are still some problems: (i)For two-dimensional (2D) directions-of-arrival (DOA) estimation, the failure in pairing causes severe performance degradation; (ii) In some practical applications, the signals received by a sensor array may come from multiple near-field sources or multiple far-field sources or their mixture. Due to different signal models for near-field and far-field sources, the existing algorithms cannot deal with them simultaneously well; and (iii) For joint azimuth and elevation direction finding, the existing estimators often encounter an estimation failure problem especially when elevation angles are between 70 and 90 degrees. In this chapter, several high-resolution methods are presented to overcome these difficulties.

In Section 1.2, a novel 2D DOA estimation algorithm without match procedure in the L-shaped array geometry is proposed. It is well known that two matched electric angles (functions of elevation and azimuth angles) must be obtained before elevation and azimuth angles are estimated. However, the failure in pairing would cause severe performance degradation. By introducing a novel electric angle, the L-shaped array configuration without any rotational invariance property between two orthogonal uniform linear sub-arrays evolves into some particular rotational invariance geometry. Thus, the steering vector is separated into two parts. One can be estimated by the rank-reduction ESPRIT algorithm and the other is obtained from the eigenvalue decomposition of one particular matrix. Finally, the elevation and azimuth angles can be easily obtained from the recovered steering vector to avoid pairing. Although it is developed for the L-shaped array configuration, the proposed algorithm can be easily extended to other array geometries such as two parallel linear sub-arrays, the rectangular array, and the symmetric circular array. In addition, the method can be used to form the rank-reduction propagator method. electric angle

In Section 1.3, a common signal model for "any-field" sources (i.e., near-field sources or far-field sources or their mixture) is given and a two-stage MUSIC algorithm is developed to localize "any-field" sources. In the first stage, one special cumulant matrix is derived and the related virtual "steering vector" is the function of the common electric angle in both near-field and far-field signal models so that DOA of near-field or far-field can be obtained from this electric angle using the conventional high-resolution MUSIC algorithm. In the second stage, another particular cumulant matrix is derived, in which the virtual "steering matrix" has full column rank no matter whether the received signals are multiple near-field sources or multiple far-field ones or their mixture. More importantly, the virtual "steering vector" can be separated into two parts, in which the first one is the function of the common electric angle in both signal models, whereas the second part is the function of the electric angle that exists only in the near-field signal model. Furthermore, by substituting the common electric angle, which is estimated in the first stage into one special Hermitian matrix formed from another MUSIC spectral function, the range of near-field sources can be obtained from the eigenvector of the Hermitian matrix. Although it is developed for azimuth angle (and range) estimation only, it can be developed further for the joint azimuth and elevation angles (as well as range) estimation.

In Section 1.4, a novel high-accuracy estimator for elevation angle is developed to avoid the estimation failure problem encountered in the conventional elevation estimators. Firstly, three cumulant matrices are constructed using fourth-order cumulants of some properly chosen array outputs of a specially designed volume array to increase the array aperture. Secondly, a parallel factor (PARAFAC) model of cumulant matrices in the cumulant domain is formed to avoid pairing parameters. Finally, a flexible and high-resolution elevation angle estimator is derived from multiple electric angles, which are solved from the above steps.

2. 2D DOA estimation without match procedure

Estimation of 2-D DOA is a key issue in sensor array signal processing such as radar, sonar, radio astronomy, and mobile communication systems [1-4]. Similar to other array geometries such as the parallel uniform linear array, the rectangular array and the circular array, there is an un-avoidable parameter association problem in the L-shaped array configuration because the failure in pairing would cause severe performance degradation. This section will give a novel 2-D DOA estimation algorithm, which does not require match procedure.

2.1 Description of the proposed algorithm

Let's consider an L-shaped sensor array with $2M+1$ omni-directional sensors, as shown in Fig. 1. The element placed at the origin is set for the referencing point. The array in the $x-z$ plane consists of two uniform linear sub-arrays with element spacing d, each being composed of M elements. Assume that L far-field, no-coherent, narrowband sources impinging on this antenna array. Let α_l and β_l be the elevation and azimuth angles of the l-th source, and thus the wave vector $\boldsymbol{\kappa}_l$ containing DOA information can be defined as $\boldsymbol{\kappa}_l = [\sin\alpha_l \cos\beta_l, \sin\alpha_l \sin\beta_l, \cos\alpha_l]$, $l = 1,\cdots,L$. After being sampled, the signals received by the sensor array can be expressed as

$$\mathbf{r}(k) = \mathbf{A}\mathbf{s}(k) + \mathbf{n}(k),\ k = 0,\cdots,K-1, \tag{1}$$

where $\mathbf{r}(k) = \left[r_{M,0}(k)\ r_{M-1,0}(k)\ \cdots\ r_{1,0}(k)\ r_{0,0}(k)\ r_{0,M}(k)\ r_{0,M-1}(k)\ \cdots\ r_{0,1}(k)\right]^T$

$$\mathbf{A}=\left[\mathbf{a}(\gamma_1,\phi_1)\cdots\mathbf{a}(\gamma_l,\phi_l)\cdots\mathbf{a}(\gamma_L,\phi_L)\right]$$

$$\mathbf{s}(k)=\left[s_1(k),\cdots,s_l(k),\cdots,s_L(k)\right]^T,\ \mathbf{a}(\gamma_l,\phi_l)=\left[e^{jM\gamma_l}\ e^{j(M-1)\gamma_l}\ \cdots e^{j\gamma_l}\ 1\ e^{jM\phi_l}\ e^{j(M-1)\phi_l}\cdots e^{j\phi_l}\right]^T$$

$$\gamma_l=-2\pi d\sin\alpha_l\cos\beta_l/\lambda$$

$$\phi_l=-2\pi d\cos\alpha_l/\lambda$$

$$\mathbf{n}(k)=\left[n_{M,0}(k)\ n_{M-1,0}(k)\ \cdots\ n_{1,0}(k)\ n_{0,0}(k)\ n_{0,M}(k)\ n_{0,M-1}(k)\ \cdots\ n_{0,1}(k)\right]^T.$$

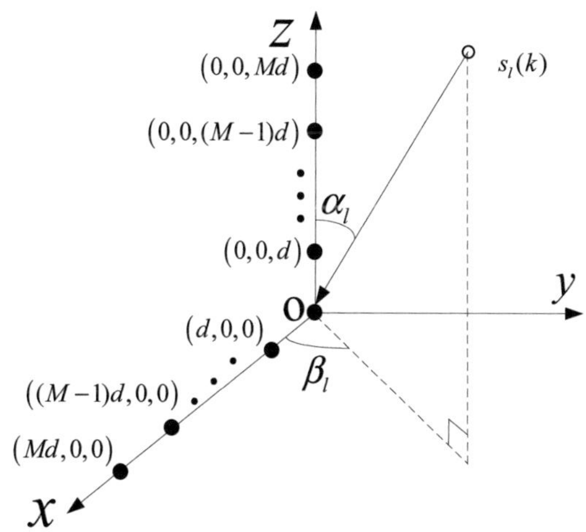

Fig. 1. L-shaped sensor array configuration

The auto-correlation matrix of $\mathbf{r}(k)$ can be expressed as $\mathbf{R}=E[\mathbf{r}(k)\mathbf{r}^H(k)]$, $=\mathbf{A}\mathbf{R}_s\mathbf{A}^H+\sigma_n^2\mathbf{I}_{2M+1}$, where $\mathbf{R}_s=E[\mathbf{s}(k)\mathbf{s}^H(k)]$, and its eigen-value decomposition (EVD) yields

$$\mathbf{R}=\mathbf{U}\mathbf{V}\mathbf{U}^H=\mathbf{U}_s\mathbf{V}_s\mathbf{U}_s^H+\mathbf{U}_n\mathbf{V}_n\mathbf{U}_n^H=[\mathbf{u}_1,\cdots,\mathbf{u}_{2M+1}]\mathrm{diag}[v_1,\cdots,v_{2M+1}][\mathbf{u}_1,\cdots,\mathbf{u}_{2M+1}]^H, \quad (2)$$

where $\mathbf{V}$ is the diagonal matrix with the eigen-values arranged as $v_1\geq\cdots\geq v_L>v_{L+1}\geq\cdots\geq v_{2M+1}$, the diagonal matrix $\mathbf{V}_n\in R^{(2M+1-L)\times(2M+1-L)}$ is composed of eigen-values $v_{L+1},v_{L+2},\cdots,v_{2M+1}$; $\mathbf{U}_n\in C^{(2M+1)\times(2M+1-L)}$ consists of the eigenvectors related to $v_{L+1},v_{L+2},\cdots,v_{2M+1}$, spanning the noise subspace of $\mathbf{R}$.

Let's define $e^{j\theta_l}=e^{j(\phi_l-\gamma_l)}$, and thus the steering vector be written in another form as:

$$\mathbf{a}(\gamma_l,\phi_l)=\left[e^{jM\gamma_l}\ e^{j(M-1)\gamma_l}\ \cdots e^{j\gamma_l}\ 1\ e^{jM\theta_l}\times e^{jM\gamma_l}\ \ e^{j(M-1)\theta_l}\times e^{j(M-1)\gamma_l}\ \cdots\ e^{j\theta_l}\times e^{j\gamma_l}\right]^T \quad (3)$$

Furthermore, $\mathbf{a}(\gamma,\phi)$ can be separated into two parts, i.e. $\mathbf{a}(\gamma,\phi)=\mathbf{a}_1\left(e^{j\theta}\right)\mathbf{a}_2\left(e^{j\gamma}\right)$, where $(M+1)$-dimensional vector $\mathbf{a}_2\left(e^{j\gamma}\right)=\left[e^{jM\gamma}\ e^{j(M-1)\gamma}\ \cdots\ e^{j\gamma}\ 1\right]^T$ and $(2M+1)\times(M+1)$ - dimensional matrix

$$\mathbf{a}_1(e^{j\theta}) = \begin{bmatrix} \mathbf{I}_{(M+1)\times(M+1)} \\ \mathrm{diag}\left\{e^{jM\theta},\cdots,e^{j\theta}\right\} \ \mathbf{0}_{M\times 1} \end{bmatrix} \tag{4}$$

Based on the theory that the noise subspace $\mathbf{U}_n$ is orthogonal to the range space of $\mathbf{A}$, $\mathbf{U}_n^H\mathbf{a}(\gamma_l,\phi_l)=\mathbf{0}$, $l=1,\cdots,L$, the electric angle pair $\{\gamma_l,\phi_l\}$, $l=1,\cdots,L$ can be found from the L deepest minima of the following MUSIC spectral function:

$$\begin{aligned} f_1(\gamma,\phi) &= \mathbf{a}^H(\gamma_l,\phi_l)\mathbf{U}_n\mathbf{U}_n^H\mathbf{a}(\gamma_l,\phi_l) = \mathbf{a}_2^H\left(e^{j\gamma_l}\right)\mathbf{a}_1^H\left(e^{j\theta_l}\right)\mathbf{U}_n\mathbf{U}_n^H\mathbf{a}_1\left(e^{j\theta_l}\right)\mathbf{a}_2\left(e^{j\gamma_l}\right) \\ &= \mathbf{a}_2^H\left(e^{j\gamma_l}\right)\mathbf{C}\left(e^{j\theta_l}\right)\mathbf{a}_2\left(e^{j\gamma_l}\right) = 0 \end{aligned} \tag{5}$$

where

$$\mathbf{C}(e^{j\theta}) = \mathbf{a}_1^H\left(e^{j\theta}\right)\mathbf{U}_n\mathbf{U}_n^H\mathbf{a}_1\left(e^{j\theta}\right) \tag{6}$$

is an $(M+1)\times(M+1)$ -dimensional Hermitian matrix.

Note that $\mathbf{a}_1\left(e^{j\theta_l}\right)\neq\mathbf{0}$ and $\mathbf{a}_2^H\left(e^{j\gamma_l}\right)\mathbf{C}(e^{j\theta_l})\mathbf{a}_2\left(e^{j\gamma_l}\right)=0$, $l=1,2,\cdots,L$. From Eq. (5)-(6), it can be seen that if and only if $e^{j\theta}=e^{j\theta_l}$, the matrix $\mathbf{C}(e^{j\theta})$ drops rank, or equivalently, when the polynomial of $x=e^{j\theta}$, $f_2(x)=\det\{\mathbf{C}(x)\}=0$. Obviously, $\hat{x}$, lying inside the unit circle and being closest to the unit circle, is actually the signal root.

Eq. (5) implies that by substituting the estimated $e^{j\hat{\theta}_l}$ into $\mathbf{C}(e^{j\theta})$ in Eq. (5), $\hat{\gamma}_l$ can be found from the minima of the following function:

$$\hat{\gamma}_l = \min_{\gamma}\ \mathbf{a}_2^H\left(e^{j\gamma}\right)\mathbf{C}(e^{j\hat{\theta}_l})\mathbf{a}_2\left(e^{j\gamma}\right) \tag{7}$$

the minima of which indicates estimation.

When $\phi_p-\gamma_p\neq\phi_q-\gamma_q+2h\pi$, $h\in\{-1,0,1\}$, $p,q\in\{1,\cdots,L\}$, **i.e.,** $\theta_p\neq\theta_q$, Eq. (7) implies that $\mathbf{a}_2\left(e^{j\hat{\gamma}_p}\right)$ is just the unique eigenvector corresponding to the smallest eigen-value of $\mathbf{C}(e^{j\hat{\theta}_p})$. However, when $\phi_p-\gamma_p=\phi_q-\gamma_q+2h\pi$, $\mathbf{a}_2\left(e^{j\hat{\gamma}_p}\right)$ is no longer the unique eigenvector corresponding to the smallest eigen-value of $\mathbf{C}(e^{j\hat{\theta}_p})$. The eigen-value decomposition (EVD) of $\mathbf{C}(e^{j\hat{\theta}_p})$ yields $\mathbf{C}(e^{j\hat{\theta}_p})=[\tilde{\mathbf{u}}_1,\cdots,\tilde{\mathbf{u}}_{M+1}]\mathrm{diag}[\tilde{v}_1,\cdots,\tilde{v}_{M+1}][\tilde{\mathbf{u}}_1,\cdots,\tilde{\mathbf{u}}_{M+1}]^{-1}$, where the eigen-values are arranged as $|v_1|\geq|v_2|\geq\cdots>|v_M|=|v_{M+1}|$. It is obvious that under the case $\phi_p-\gamma_p=\phi_q-\gamma_q+2h\pi$ $\mathbf{a}_2\left(e^{j\hat{\gamma}_p}\right)$ and $\mathbf{a}_2\left(e^{j\hat{\gamma}_q}\right)$ are the linear combinations of two eigenvectors $\{\tilde{\mathbf{u}}_M,\tilde{\mathbf{u}}_{M+1}\}$, which are orthogonal to $\tilde{\mathbf{U}}_s=\{\tilde{\mathbf{u}}_1,\tilde{\mathbf{u}}_2,\cdots,\tilde{\mathbf{u}}_{M-1}\}$. Obviously, both $e^{j\hat{\gamma}_p}$ and $e^{j\hat{\gamma}_q}$ are the roots of $f_3(x)=\mathbf{a}_2^H(x)\tilde{\mathbf{U}}_s\tilde{\mathbf{U}}_s^H\mathbf{a}_2(x)=0$.

From the estimates $\left\{e^{j\hat{\gamma}_l},e^{j\hat{\theta}_l}\right\}$, the elevation and azimuth angle estimates can be given as $\hat{\alpha}_l=\arccos\left(-\angle\left(e^{j\hat{\theta}_l}\times e^{j\hat{\gamma}_l}\right)\lambda/(2\pi d)\right)$ and $\hat{\beta}_l=\arccos\left(-\left(\angle e^{j\hat{\gamma}_l}\right)\lambda/(2\pi d\sin\hat{\alpha}_l)\right)$, respectively. Since $\mathbf{a}_2\left(e^{j\hat{\gamma}_l}\right)$ is related to $\mathbf{C}(e^{j\hat{\theta}_l})$ (i.e., corresponding to $\mathbf{a}_1\left(e^{j\hat{\theta}_l}\right)$), the proposed algorithm can avoid pairing parameters. In addition, it avoids the spectral search because both $e^{j\hat{\gamma}_l}$ and $e^{j\hat{\theta}_l}$ are estimated by solving polynomial roots.

2.2 Simulation results

To verify the effectiveness of the proposed algorithm, let's consider an L-shaped array with 13 elements as shown in Fig.1. These sensor locations are in unit of $d = \lambda / 2$. Two uncorrelated equivalent-power sources ($e^{j0.2\pi k}$ and $e^{j0.25\pi k}$), respectively with DOAs of $(\alpha_1 = 60^o, \beta_1 = 35^o)$ and $(\alpha_2 = 40^o, \beta_2 = 55^o)$, impinge on this array. The root mean square error (RMSE) is used as the performance measure. All results provided are based on 500 independent runs. The RMSE for DOA estimation is defined as

$$\text{RMSE(the } l\text{th signal)} = \sqrt{\frac{1}{500}\sum_{i=1}^{500}\left(\hat{\alpha}_{i,l} - \alpha_l\right)^2} \quad (8)$$

in which $\hat{\alpha}_{i,l}$ (in unit of degree) stands for the estimation of the l-th elevation α_l in the i-th trial. For comparison, the propagator method [6, 9] and the ESPRIT method [8, 11] with correct pairing are simultaneously executed.

In the first experiment, the effect of signal-to-noise (SNR) on the performance of the proposed algorithm is investigated. The number of snapshots is set to 400 and the SNR varies from 0 to 30 dB. The averaged performances (RMSE of elevation and azimuth angle estimations versus SNR for two sources) over 500 Monte Carlo runs are shown in Figs. 2 and 3. As expected, when the SNR increases, the RMSE of the estimated parameters decrease. In addition, it is observed that the proposed algorithm improves the performance slightly compared to the conventional ESPRIT algorithm, which must have a precise association procedure.

In the second experiment, the influence of snapshot number on the performance of the proposed algorithm is explored. The same parameters as that of the second experiment are used, except that the SNR is fixed at 10 dB and the number of snapshots varies from 200 to 2000. The averaged performances (RMSE of elevation and azimuth angle estimations versus snapshot number for two sources) over 500 Monte Carlo runs are shown in Figs. 4 and 5. From these figures, it can be seen that RMSE of the elevation and azimuth estimations decrease as snapshot number increases. In addition, the proposed algorithm has higher estimation accuracy than the ESPRIT method.

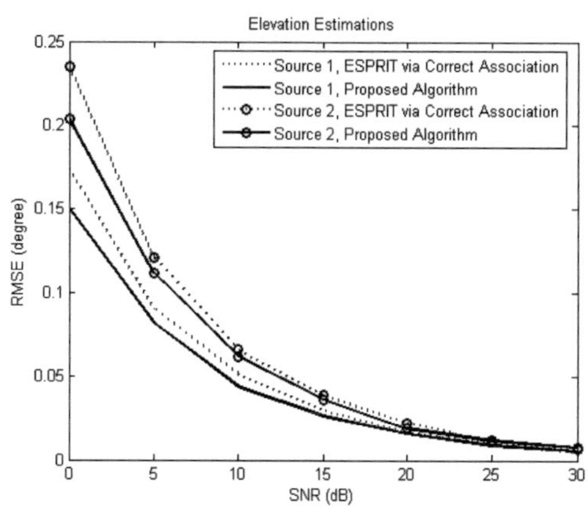

Fig. 2. RMSE of elevation angle estimations versus SNR

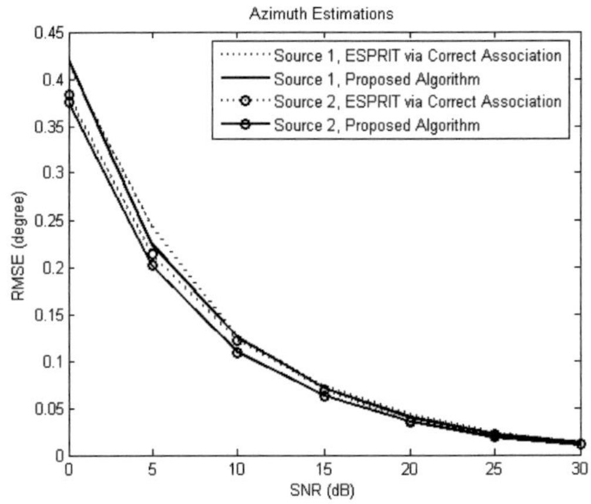

Fig. 3. RMSE of azimuth angle estimations versus SNR

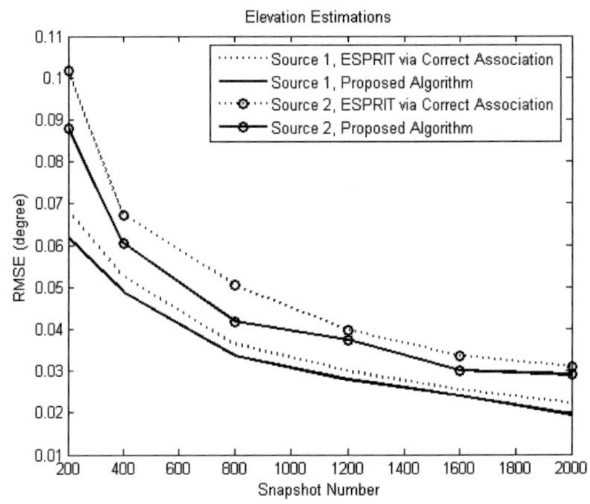

Fig. 4. RMSE of elevation angle estimations versus snapshot number

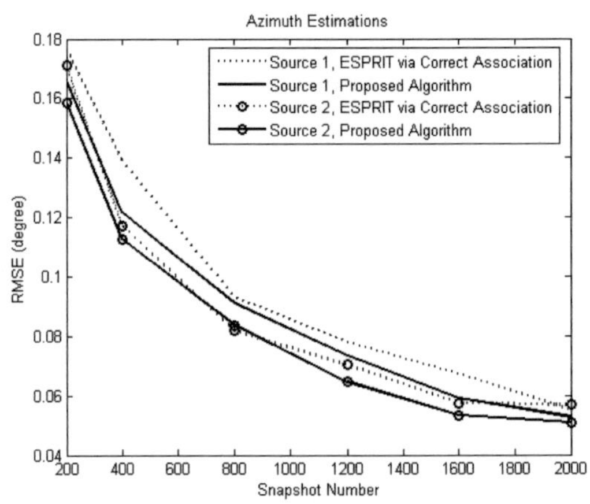

Fig. 5. RMSE of azimuth angle estimations versus snapshot number

From the above experiments, it can be seen that (i) since $\mathbf{a}_2\left(e^{j\hat{\gamma}_l}\right)$ is related to $\mathbf{C}(e^{j\hat{\theta}_l})$ (i.e. corresponding to $\mathbf{a}_1\left(e^{j\hat{\theta}_l}\right)$), the proposed algorithm can avoid pairing parameters; and (ii) the proposed algorithm avoids the spectral search due to that both $e^{j\hat{\gamma}_l}$ and $e^{j\hat{\theta}_l}$ are estimated by solving polynomial roots.

3. Passive localization of mixed near-field and far-field sources

In some practical applications, the signals received by an array are often the mixture of near-field and far-field sources, such as speaker localization using microphone arrays and guidance (homing) systems [12-19]. For example, in the application of speaker localization using microphone arrays, each speaker may be in the near-field or far-field of the array [16]. In this case, either existing near-field source localization methods or far-field source those may fail in localizing mixed near-field and far-field sources. This section will give a new passive source localization algorithm, which can localize near-field sources or far-field sources or their mixture.

3.1 Description of the proposed algorithm

Consider that L (near-field[1] or far-field) narrowband, independent radiating sources, impinge on the uniform linear array (ULA) with $2N+1$ elements as shown in Fig.6. Let the 0 th sensor be the phase reference point. After sampled with a proper rate that satisfies the Nyquist rate, the signal received by the i th sensor can be expressed as [5-11]

$$x_i(k)=\sum_{l=1}^{L}s_l(k)e^{j\tau_{il}}+n_i(k)\ ,\ -N\le i\le N\,,\ k=0,\cdots,K-1\,, \tag{9}$$

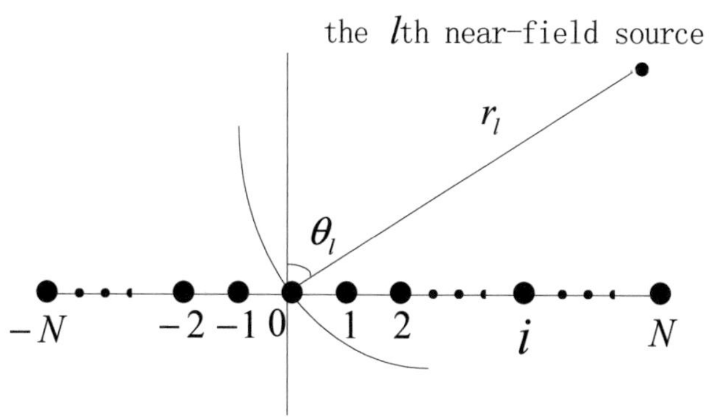

Fig. 6. Uniform linear array configuration

[1] Note that *Fresnel* zone (i.e. near-field) lies in the radiating zone $[\frac{1}{2\pi}\lambda,\frac{1}{\lambda}2D^2]$, where λ and D are signal wavelength and array dimension, respectively (see [4] for details). Whereas far-field means the radiating zone beyond $[0,\frac{1}{\lambda}2D^2]$.

Where K is the snapshot number, $s_l(k)$ is the l th narrowband source, $n_i(k)$ is the additive Gaussian noise. In addition, τ_{il} is the delay associated with the l th source propagation time between the 0 th and i th sensors. If the l th source is near-field one (the azimuth DOA θ_l and the range r_l), $\tau_{il} = i\gamma_l + i^2\phi_l$, where γ_l and ϕ_l are called electric and given by

$$\gamma_l = -2\pi\frac{d}{\lambda}\sin(\theta_l)\text{, and } \phi_l = \pi\frac{d^2}{\lambda r_l}\cos^2(\theta_l).$$

Otherwise, if the l th source is far-field one, $\tau_{il} = i\gamma_l$ [2, 3], where ϕ_l is approximated by zero due to father range of far-field source. Therefore, a far-field source can be considered as the special near-field one, where $\phi_l = 0$.

In a matrix form, Eq. (1) can be written as

$$\mathbf{x}(k) = \mathbf{A}\mathbf{s}(k) + \mathbf{n}(k) \tag{10}$$

$$\text{where } \mathbf{A} = [\mathbf{a}(\gamma_1,\phi_1)\cdots\mathbf{a}(\gamma_l,\phi_l)\cdots\mathbf{a}(\gamma_L,\phi_L)]$$

$$\mathbf{n}(k) = [n_{-N}(k),\cdots,n_0(k),n_1(k),\cdots,n_N(k)]^T,$$

$$\mathbf{s}(k) = [s_1(k),\cdots,s_l(k),\cdots,s_L(k)]^T$$

$$\mathbf{x}(k) = [x_{-N}(k),\cdots,x_0(k),x_1(k),\cdots,x_N(k)]^T.$$

Note that the form of steering vector $\mathbf{a}(\gamma_l,\phi_l)$ depends on whether the l th source is far-field one or near-field one. If this source is near-field one,

$$\mathbf{a}(\gamma_l,\phi_l) = \left[e^{j[(-N)\gamma_l+(-N)^2\phi_l]}\ e^{j[(-N+1)\gamma_l+(-N+1)^2\phi_l]}\ \cdots\ e^{j[N\gamma_l+N^2\phi_l]}\right]^T.$$

Otherwise, if this source lies in the far field,

$$\mathbf{a}(\gamma_l,\phi_l) = \left[e^{j[(-N)\gamma_l]}\ e^{j[(-N+1)\gamma_l]}\ \cdots\ e^{j[N\gamma_l]}\right]^T.$$

Let's begin with the fourth-order cumulant of the sensor outputs, which can be expressed as

$$\begin{aligned}
&\text{cum}\{x_m(k),x_n^*(k),x_p^*(k),x_q(k)\}\\
&= \text{cum}\{\sum_{l=1}^{L}s_l(k)e^{j(m\gamma_l+m^2\phi_l)},\left(\sum_{l=1}^{L}s_l(k)e^{j(n\gamma_l+n^2\phi_l)}\right)^*,\left(\sum_{l=1}^{L}s_l(k)e^{j(p\gamma_l+p^2\phi_l)}\right)^*,\sum_{l=1}^{L}s_l(k)e^{j(q\gamma_l+q^2\phi_l)}\}\\
&= \sum_{l=1}^{L}e^{j\{[(m-n)-(p-q)]\gamma_l+[(m^2-n^2)-(p^2-q^2)]\phi_l\}}\text{cum}\{s_l(t),s_l^*(t),s_l^*(t),s_l(t)\}\\
&= \sum_{l=1}^{L}c_{4,sl}e^{j\{[(m-n)-(p-q)]\gamma_l+[(m^2-n^2)-(p^2-q^2)]\phi_l\}}
\end{aligned} \tag{11}$$

$$m,n,p,q \in [-N,N]$$

where $c_{4,sl} = \text{cum}\{s_l(t), s_l^*(t), s_l^*(t), s_l(t)\}$ is the kurtosis of the l th signal, and the superscript * denotes the complex conjugate.

To construct a matrix with full rank for arbitrary-field sources , let $n = -m$ and $q = 0$. Thus, Eq. (11) becomes

$$\text{cum}\{x_m(k), x_{-m}^*(k), x_p^*(k), x_q(k)\} = \sum_{l=1}^{L} c_{4,sl} e^{j2m\gamma_l} \left(e^{j(p\gamma_l + p^2\phi_l)}\right)^*, \quad m,p \in [-N,N] \tag{12}$$

Let $\bar{m} = m + N + 1$ and $\bar{p} = p + N + 1$, and thus $\bar{m}, \bar{p} \in [1, 2N+1]$. Based on the idea from (11)-(12), a special $(2N+1)\times(2N+1)$-dimensional cumulant matrix $\mathbf{C}$ can be defined, the $(\bar{m},\bar{p})$ th element of which can be given by

$$\begin{aligned} \mathbf{C}(\bar{m},\bar{p}) &= \text{cum}\{x_{\bar{m}-N-1}(k), x_{-\bar{m}+N+1}^*(k), x_{\bar{p}-N-1}^*(k), x_0(k)\} \\ &= \sum_{l=1}^{L} c_{4,sl} e^{j2(\bar{m}-N-1)\gamma_l} \left(e^{j(\bar{p}-N-1)\gamma_l + j(\bar{p}-N-1)^2\phi_l}\right)^* \qquad \bar{m},\bar{p} \in [1, 2N+1] \end{aligned} \tag{13}$$

Note that the $(2N+1)\times(2N+1)$ matrix $\mathbf{C}$ can be represented in a compact matrix form as $\mathbf{C} = \mathbf{B}\mathbf{C}_{4s}\mathbf{A}^H$, where the superscript H stands for the Hermitian transpose, $\mathbf{C}_{4s} = \text{diag}[c_{4,s1}, \cdots, c_{4,sl}, \cdots, c_{4,sL}]$, virtual "steering matrix" $\mathbf{B} = [\mathbf{b}(\gamma_1), \cdots, \mathbf{b}(\gamma_l), \cdots, \mathbf{b}(\gamma_L)]$, and virtual "steering vector" $\mathbf{b}(\gamma_l) = [e^{-j2N\gamma_l}, e^{-j(2N-2)\gamma_l}, \cdots 1, \cdots, e^{j2N\gamma_l}]^T$, $l = 1, \cdots, L$.

The singular value decomposition (SVD) of $\mathbf{C}$ yields

$$\mathbf{C} = \mathbf{B}\mathbf{C}_{4s}\mathbf{A}^H = \mathbf{W}\mathbf{\Sigma}\mathbf{Z}^H = [\mathbf{w}_1, \cdots, \mathbf{w}_{2N+1}]\text{diag}(\sigma_1, \cdots, \sigma_{2N+1})[\mathbf{z}_1, \cdots, \mathbf{z}_{2N+1}]^H \tag{14}$$

where $\mathbf{\Sigma}$ is the diagonal matrix with the singular values arranged as $|\sigma_1| \geq \cdots \geq |\sigma_L| > |\sigma_{L+1}| \geq \cdots \geq |\sigma_{2N+1}|$. Let $\mathbf{W}_s \in C^{(2N+1)\times L}$, which spans the signal subspace of $\mathbf{B}$, consists of the left singular vectors $\mathbf{w}_1, \mathbf{w}_2, \cdots, \mathbf{w}_L$. Similarly, $\mathbf{Z}_n \in C^{(2N+1)\times(2N+1-L)}$, which is orthogonal to $\mathbf{A}$, consists of the right singular vectors $\mathbf{z}_{L+1}, \mathbf{z}_{L+2}, \cdots, \mathbf{z}_{2N+1}$.

Based on the first $2N$ lines $\mathbf{W}_{s1}$ and last $2N$ lines $\mathbf{W}_{s2}$, $\gamma_l, l = 1, \cdots, L$ can be easily estimated from the eigen-values of the following matrix [3] :

$$\gamma_l = \angle(\mathbf{\Phi}(l,l)) / 2 \tag{15}$$

where $\mathbf{\Phi} = \text{diag}[e^{j2\gamma_1}, \cdots, e^{j2\gamma_l}, \cdots, e^{j2\gamma_L}]$ is the eigen-value matrix $\mathbf{W}_{s1}^{\#}\mathbf{W}_{s2}$, i.e. $\mathbf{W}_{s1}^{\#}\mathbf{W}_{s2} = \mathbf{T}\mathbf{\Phi}\mathbf{T}^{-1}$.

By substituting the estimate $\hat{\gamma}_l$ into $\mathbf{a}(\gamma, \phi)$, the minima of the following function can be found.

$$\hat{\phi}_l = \min_{\phi} \mathbf{a}^H(\hat{\gamma}_l, \phi)\mathbf{Z}_n\mathbf{Z}_n^H\mathbf{a}(\hat{\gamma}_l, \phi) = \min_{\phi} \mathbf{a}_2^H(\phi)\mathbf{a}_1^H(\hat{\gamma}_l)\mathbf{Z}_n\mathbf{Z}_n^H\mathbf{a}_1(\hat{\gamma}_l)\mathbf{a}_2(\phi), l = 1, \cdots, L$$

where $\mathbf{a}_1(\gamma) = \begin{bmatrix} \text{diag}\{e^{-jN\gamma}, \cdots, e^{-j\gamma}, 1\} \\ \text{diag}^{\text{anti}}\{e^{j\gamma}, \cdots, e^{jN\gamma}\} \ \ \mathbf{0}_{N\times 1} \end{bmatrix}$, $\mathbf{a}_2(\phi) = \begin{bmatrix} e^{jN^2\phi} & e^{j(N-1)^2\phi} & \cdots & e^{j\phi} & 1 \end{bmatrix}^T$.

Eq. (16) implies that $\mathbf{a}_2\left(\hat{\phi}_l\right)$ is just the eigenvector corresponding to the smallest eigen-value of $\mathbf{a}_1^H(\hat{\gamma}_l)\mathbf{Z}_n\mathbf{Z}_n^H\mathbf{a}_1(\hat{\gamma}_l)$, and $\hat{\phi}_l$ is easily solved from $\mathbf{a}_2\left(\hat{\phi}_l\right)$.

The DOA and range estimates of the l th source can be in turn expressed as:

$$\hat{\theta}_l = \arcsin\left(-\frac{\hat{\gamma}_l\lambda}{2\pi d}\right) \tag{17}$$

and

$$\hat{r}_l = \frac{\pi d^2}{\lambda\hat{\phi}_l}\cos^2(\hat{\theta}_l)\,,\ l = 1,\cdots,L\,. \tag{18}$$

In fact, if the l th source is far-field one, the estimate $\hat{\phi}_l$ would approach to zero. Thus, whether the l th source is near-field or far-field one can be determined. Since both $\mathbf{B}$ and $\mathbf{A}$ are of full column rank no matter whether the received signals be pure far-field sources or pure near-field sources or mixed far-field and near-field sources, the proposed algorithm can deal with arbitrary-field sources well.

3.2 Description of the proposed algorithm

Some simulations are conducted in this section to assess the ability of the proposed algorithm to localize near-field, far-field, as well as mixed near-field and far-field sources.

Two near-field sources are located at $\{\theta_1 = 10^o,\ r_1 = 0.5\lambda\}$ (i.e. $\{\gamma_1 = -0.2728, \phi_1\ \ 0.3809\}$) and $\{\theta_2 = 20^o,\ r_2 = 1.0\lambda\}$ (i.e. $\{\gamma_2 = -0.5372, \phi_2\ \ 0.1734\}$), respectively. The snapshot number and SNR are fixed at 400 and 10 dB. The scatter plot of estimated $(\hat{\gamma}_l, \hat{\phi}_l)$ pairs from 500 independent trials using the proposed algorithm, the near-field source localization algorithm (i.e. ESPRIT), and the far-field source localization algorithm (i.e. MUSIC) are shown in Figs. 7-9, respectively. From these figures, it can be seen that the far-field source localization algorithm fails in localizing near-field sources.

The near-field source is located at $\{\theta_1 = 10^o,\ r_1 = 0.5\lambda\}$ (i.e. $\{\gamma_1 = -0.2728, \phi_1\ \ 0.3809\}$); whereas the far-field source is localized at $\{\theta_2 = 20^o, r_2 = +\infty\}$ (i.e. $\{\gamma_2 = -0.5372, \phi_2\ \ 0\}$). The snapshot number and SNR are fixed at 400 and 10 dB, respectively. The scatter plot of estimated $(\hat{\gamma}_l, \hat{\phi}_l)$ pairs from 500 independent trials using the proposed algorithm, the near-field source localization algorithm (ESPRIT), and the far-field source localization algorithm (MUSIC) are shown in Figs. 10-12, respectively. These figures show that the far-field source localization algorithm (MUSIC) fails in estimating azimuth DOA of the near-field source. However, the proposed algorithm performs well in localizing both near-field and far-field sources.

Two far-field sources are localized at $\{\theta_1 = 10^o, r_1 = +\infty\}$ (i.e. $\{\gamma_1 = -0.2728, \phi_1\ \ 0\}$) and $\{\theta_2 = 20^o, r_2 = +\infty\}$ (i.e. $\{\gamma_2 = -0.5372, \phi_2\ \ 0\}$), respectively. When the snapshot number and SNR are fixed respectively at 400 and 10 dB, the scatter plot of estimated $(\hat{\gamma}_l, \hat{\phi}_l)$ pairs from 500 independent trials using the proposed algorithm, the near-field source localization algorithm (ESPRIT), and the far-field source localization algorithm (MUSIC) are shown in Figs. 13-15, respectively. From these figures, it can be seen that the near-field source localization algorithm fails in localizing far-field sources, but the proposed algorithm performs well in estimating azimuth DOA of the two far-field source.

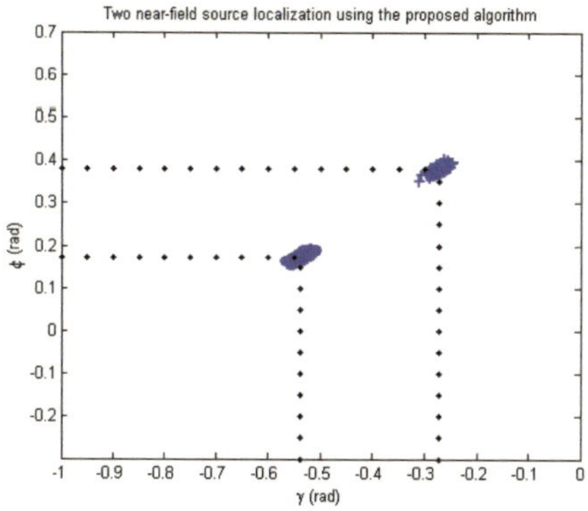

Fig. 7. Scatter plot of estimated (γ,ϕ) pairs for two near-field sources using the proposed algorithm

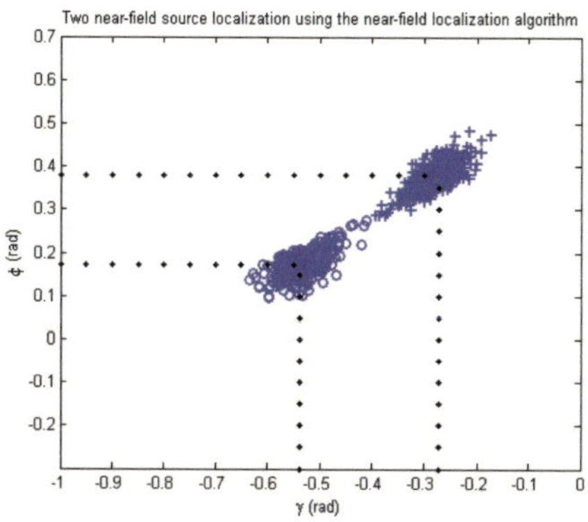

Fig. 8. Scatter plot of estimated (γ,ϕ) pairs for two near-field sources using the near-field source localization algorithm (ESPRIT)

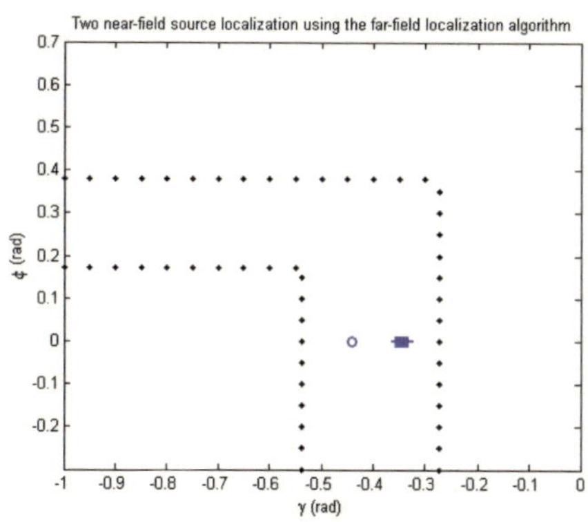

Fig. 9. Scatter plot of estimated (γ,ϕ) pairs for two near-field sources using the far-field source localization algorithm (MUSIC)

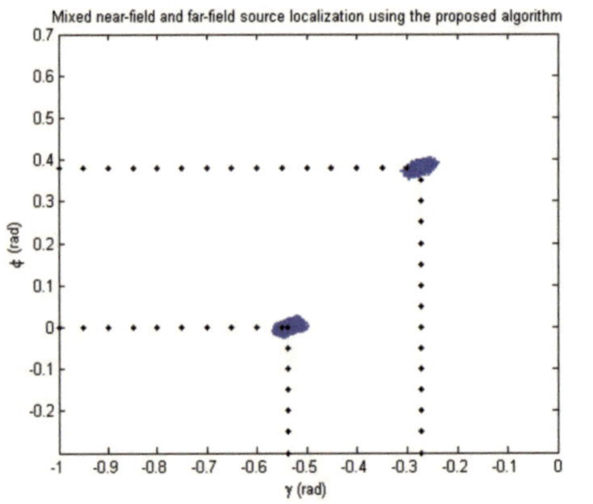

Fig. 10. Scatter plot of estimated (γ,ϕ) pairs for mixed near-field and far-field sources using the proposed algorithm

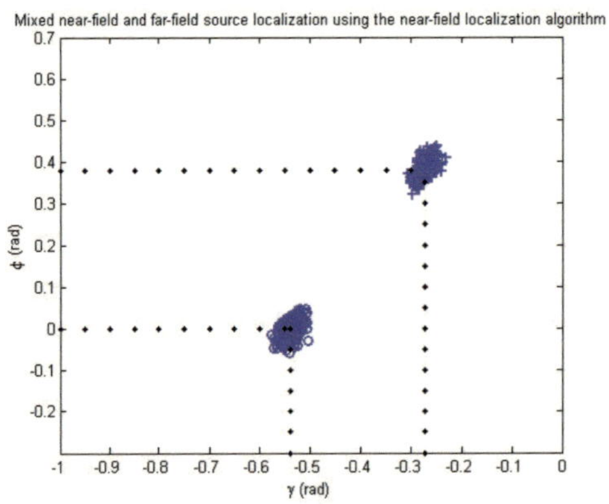

Fig. 11. Scatter plot of estimated (γ,ϕ) pairs for mixed near-field and far-field sources using the near-field source localization algorithm (ESPRIT)

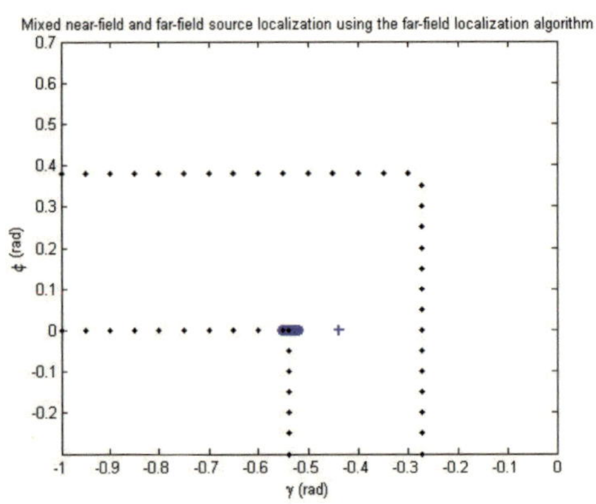

Fig. 12. Scatter plot of estimated (γ,ϕ) pairs for mixed near-field and far-field sources using the far-field source localization algorithm (MUSIC)

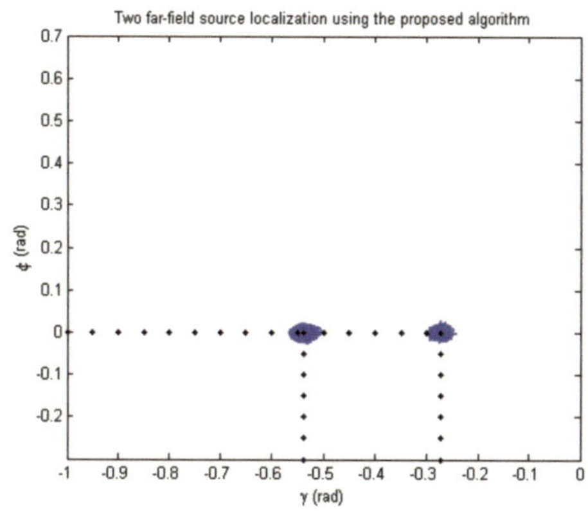

Fig. 13. Scatter plot of estimated (γ,ϕ) pairs for two far-field sources using the proposed algorithm

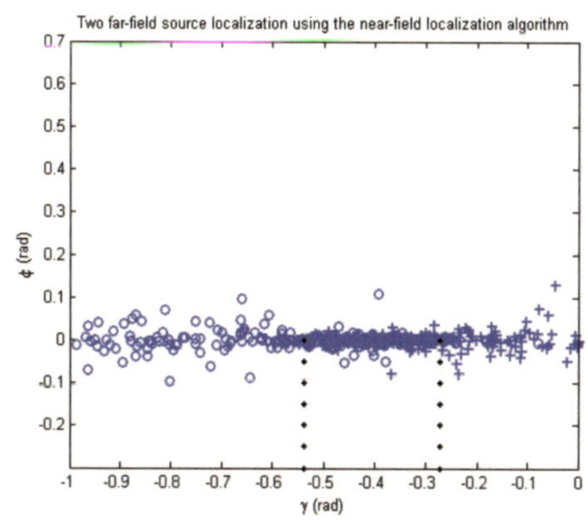

Fig. 14. Scatter plot of estimated (γ,ϕ) pairs for two far-field sources using the near-field source localization algorithm (ESPRIT)

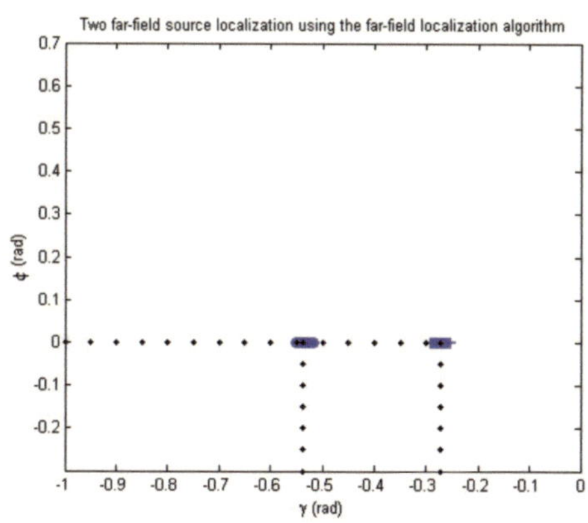

Fig. 15. Scatter plot of estimated (γ,ϕ) pairs for two far-field sources using the far-field source localization algorithm (MUSIC)

From these simulations, it can be seen that no matter whether the received signals are near-field sources, far-field sources or their mixture, the proposed algorithm can perform better in localizing these sources.

4. New estimator for elevation angle

4.1 Proposed estimator

When elevation angles are between 70 and 90 degrees, the estimator

$$\operatorname{asin}\left(\frac{\lambda}{2\pi d}\sqrt{\hat{\gamma}_l^2+\hat{\phi}_l^2}\right)$$

may fail because imperfect estimations of $(\hat{\gamma}_l, \hat{\phi}_l)$ result in

$$\frac{\lambda}{2\pi d}\sqrt{\hat{\gamma}_l^2+\hat{\phi}_l^2}$$

being greater than 1, causing the calculation of

$$\operatorname{asin}\left(\frac{\lambda}{2\pi d}\sqrt{\hat{\gamma}_l^2+\hat{\phi}_l^2}\right)$$

to fail, where $\gamma_l = -2\pi d \sin\alpha_l \cos\beta_l / \lambda$ and $\phi_l = -2\pi d \sin\alpha_l \sin\beta_l / \lambda$. On the other hand, some algorithms adopt another estimator

$$\operatorname{acos}\left(\frac{-\lambda\hat{\vartheta}_l}{2\pi d}\right),$$

which is of low estimation accuracy when elevation angles are between 0 and 20 degrees ($\vartheta_l = -2\pi d \cos\alpha_l / \lambda$). Note that $\cos(20^o) = \sin(70^o) > 0.9$ and thus it is impossible that both $\cos(\alpha)$ and $\sin(\alpha)$ are greater than 0.9 simultaneously. Therefore, e

$$\operatorname{asin}\left(\frac{\lambda}{2\pi d}\sqrt{\hat{\gamma}_l^2+\hat{\phi}_l^2}\right) \text{ and } \operatorname{acos}\left(\frac{-\lambda\hat{\vartheta}_l}{2\pi d}\right)$$

can be combined to form a new elevation angle estimator, which can efficiently avoid estimation failure and is of high estimation accuracy.

$$\alpha_l = \begin{cases} \frac{1}{2}\left\{\operatorname{acos}\left(\frac{-\lambda\hat{\vartheta}_l}{2\pi d}\right) + \operatorname{asin}\left(\frac{\lambda}{2\pi d}\sqrt{\hat{\gamma}_l^2+\hat{\phi}_l^2}\right)\right\} & \text{if } \left(\left|\frac{\lambda\hat{\vartheta}_l}{2\pi d}\right| < 0.9\right) \text{ and } \left(\left|\frac{\lambda}{2\pi d}\sqrt{\hat{\gamma}_l^2+\hat{\phi}_l^2}\right| < 0.9\right) \\ \operatorname{acos}\left(\frac{-\lambda\hat{\vartheta}_l}{2\pi d}\right) & \text{if } \left(\left|\frac{\lambda\hat{\vartheta}_l}{2\pi d}\right| < 0.9\right) \text{ and } \left(\left|\frac{\lambda}{2\pi d}\sqrt{\hat{\gamma}_l^2+\hat{\phi}_l^2}\right| \geq 0.9\right) \\ \operatorname{asin}\left(\frac{\lambda}{2\pi d}\sqrt{\hat{\gamma}_l^2+\hat{\phi}_l^2}\right) & \text{if } \left(\left|\frac{\lambda\hat{\vartheta}_l}{2\pi d}\right| \geq 0.9\right) \text{ and } \left(\left|\frac{\lambda}{2\pi d}\sqrt{\hat{\gamma}_l^2+\hat{\phi}_l^2}\right| < 0.9\right) \end{cases} \quad (19)$$

4.2 Simulation results

To access the effectiveness of the proposed elevation estimator, we consider a two L-shape arrays with 16 elements as shown in Fig. 16. These sensor locations are in unit of $d=\lambda/2$. We consider a single source case ($e^{j0.2\pi k}$): elevation angle α and azimuth angle β vary from $(0^0,0^0)$ to $(90^0,90^0)$ with 5^0 increment. The snapshot number and the SNR are set to 200 and 10dB, respectively. The received signals are polluted by zero-mean additive white Gaussian noises. We use the root mean square error (RMSE) as the performance measure. All results provided are based on 500 independent runs. For each (α,β), we conduct 500 trials, and count the estimation failure times as well as the averaged performance only from successful trials (RMSE of elevation angle estimations versus different azimuth-elevation pair for this single source). Fig. 17 and 18 give the averaged performance counted from successful trials and the corresponding Failure Rates (FR, the failure times divided by 500) of the estimator

$$\operatorname{asin}\left(\frac{\lambda}{2\pi \mathrm{d}}\sqrt{\hat{\gamma}_l^2+\hat{\phi}_l^2}\right),$$

respectively.

Since the estimator

$$\operatorname{acos}\left(\frac{-\lambda\hat{\vartheta}_l}{2\pi d}\right)$$

shows no failure for all pair angles, we only show its averaged performance (RMSE of elevation angle estimations versus different azimuth-elevation pair for this single source) in Fig. 19.

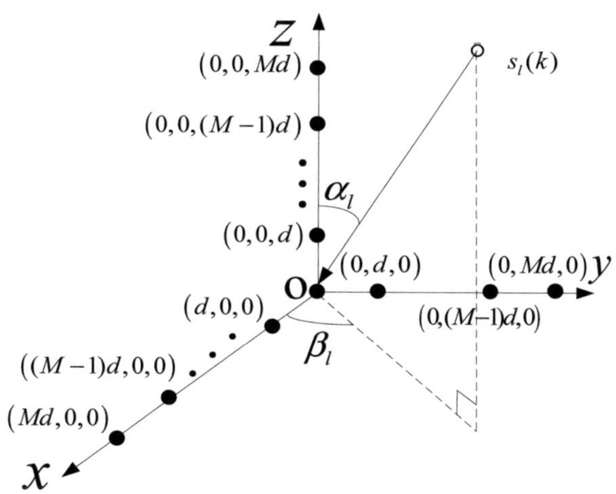

Fig. 16. Two L-shape array configuration

From Figs. 17 and 18, it can be seen that the estimator

$$\operatorname{asin}\left(\frac{\lambda}{2\pi \mathrm{d}}\sqrt{\hat{\gamma}_l^2+\hat{\phi}_l^2}\right)$$

may break downs especially when elevation angles are between 70^0 and 90^0. In addition, it is obvious that the estimator

$$\operatorname{asin}\left(\frac{\lambda}{2\pi \mathrm{d}}\sqrt{\hat{\gamma}_l^2+\hat{\phi}_l^2}\right)$$

has lower estimation accuracy in region $[70^0,90^0]$ than other regions. As the elevation angle approximates to 90^0, the failure rate increases, and the related RMSE (from the successful trails) increases [15]. Although the estimator

$$\operatorname{acos}\left(\frac{-\lambda\hat{\vartheta}_l}{2\pi d}\right)$$

shows no failure for all pair angles, we observe from Fig.19 that when the elevation angle lies in $[0^0,20^0]$, the related RMSE increases. In addition, the averaged performance of the proposed estimator (RMSE of azimuth angle estimations versus different azimuth-elevation pair for this single source) is given in Fig. 20, which shows that the proposed estimator improves the performance significantly compared to

$$\operatorname{asin}\left(\frac{\lambda}{2\pi \mathrm{d}}\sqrt{\hat{\gamma}_l^2+\hat{\phi}_l^2}\right) \text{ and } \operatorname{acos}\left(\frac{-\lambda\hat{\vartheta}_l}{2\pi d}\right)$$

especially in the regions $[0^0,20^0]$ and $[70^0,90^0]$.

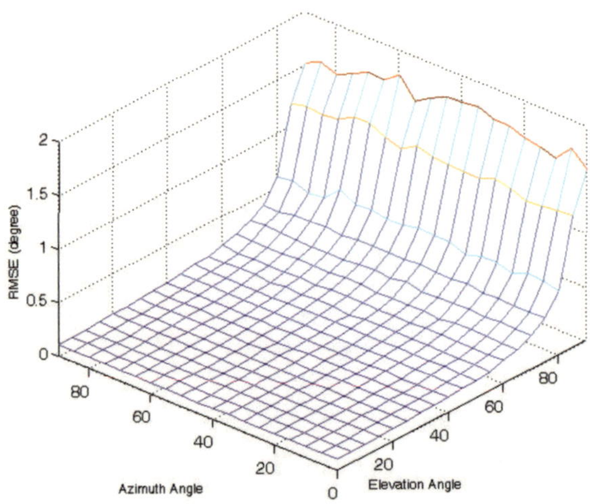

Fig. 17. Averaged performance (counted from successful trials) of elevation angle estimations versus different azimuth-elevation pair using the estimator $\operatorname{asin}\left(\frac{\lambda}{2\pi \mathrm{d}}\sqrt{\hat{\gamma}_l^2+\hat{\phi}_l^2}\right)$

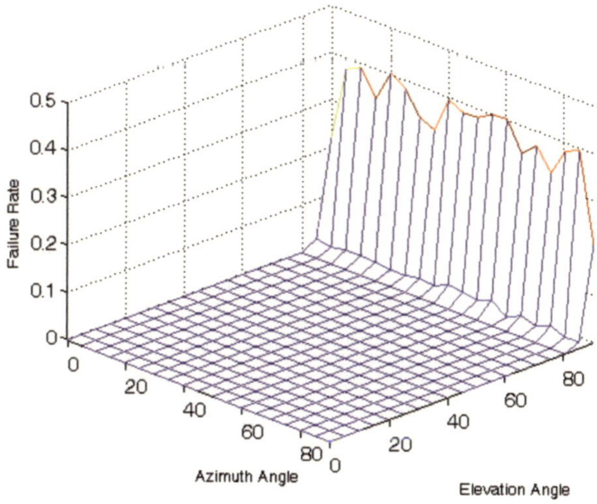

Fig. 18. Failure rates of elevation angle estimations versus different azimuth-elevation pair using the estimator $\operatorname{asin}\left(\frac{\lambda}{2\pi \mathrm{d}}\sqrt{\hat{\gamma}_l^2+\hat{\phi}_l^2}\right)$

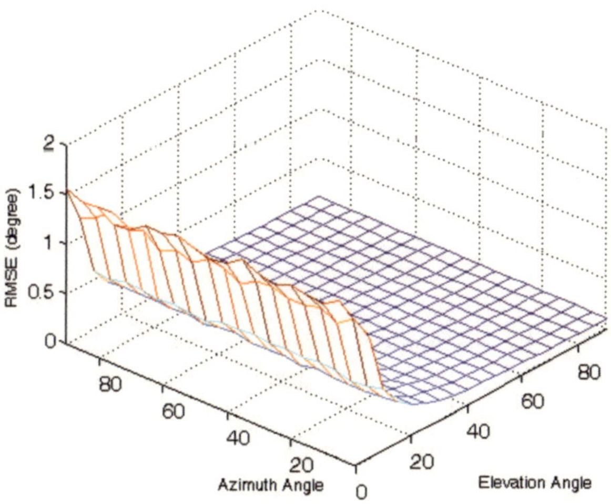

Fig. 19. Averaged performance of elevation angle estimations versus different azimuth-elevation pair using the estimator $\operatorname{acos}\left(\frac{-\lambda\hat{\vartheta}_l}{2\pi d}\right)$

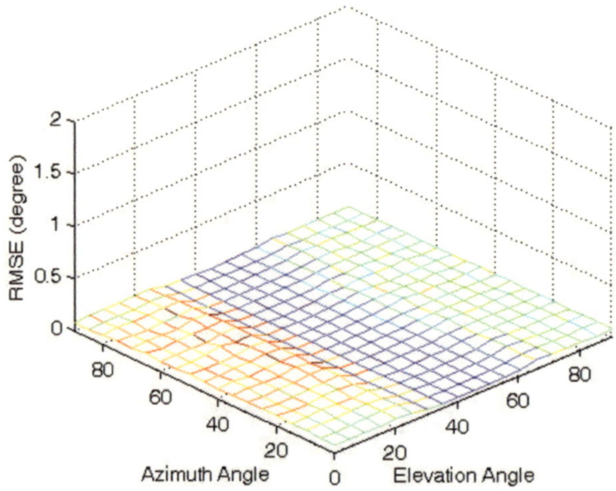

Fig. 20. Averaged performance of elevation angle estimations versus different azimuth-elevation pair using the proposed estim

From these simulations, it can be seen that the proposed estimator combines

$$\mathrm{asin}\left(\frac{\lambda}{2\pi \mathrm{d}}\sqrt{\hat{\gamma}_l^2+\hat{\phi}_l^2}\right) \text{ and } \mathrm{acos}\left(\frac{-\lambda\hat{\vartheta}_l}{2\pi d}\right),$$

and thus avoids both estimation failure and low resolution.

5. Conclusion

In this chapter, several novel high-resolution methods are introduced to overcome the difficulties encountered in the passive source localization of sensor array, i.e. pairing failure, mixed near-field and far-field source localization, and estimation failure problems. Although they have been developed for the uniform linear array and L-shaped array, these algorithms can be easily extended to other sensor array configurations.

6. Acknowledgments

This work was supported by the National Natural Science Foundation of China under Grants 61172123, 60901059 and 61075044, by China Postdoctoral Science Foundation funded project under Grant 201003679 and 20100481355, by the Natural Science Foundation under 2010JQ8001 of Shaanxi Province, and by the Discipline Union Fund under Grant 116-210905 of Xi'an University of Technology.

7. References

[1] Krim, H. & Viberg M. (1996) Two Decades of Array Signal Processing Research: the Parametric Approach. *IEEE Signal Processing Magazine,* Vol. 13, No. 4, pp. 67-94, ISSN 1053-5888

[2] Johnson, R. (1993) *Antenna Engineering Handbook*. Third Edition, New York: McGraw-Hill, Inc., ISBN 0-07-032381-X

[3] Roy, R. & Kailath, T. (1989) ESPRIT-Estimation of Signal Parameters Via Rotational Invariance Techniques. *IEEE Trans. Acoustics, Speech, and Signal Processing*, Vol. 37, No. 7, pp. 984-995, ISSN 1053-587X

[4] Schmidt, R. (1986) Multiple Emitter Location and Signal Parameter Estimation. *IEEE Trans. Antennas and Propagation*, Vol. 34, No. 3, pp. 276-280, ISSN 0018-926X

[5] Marcos, S.; Marsal, A. & Benidir, M. (1995) The Propagator Method for Source Bearing Estimation. Signal Processing, Vol. 42, No. 8, pp. 121-138, ISSN 0165-1684

[6] Hua, Y.; Sarkar, T. & Weiner, D. (1991) An L-shaped Array for Estimation 2-D Directions of Wave Arrival. *IEEE Trans. Antennas and Propagation*, Vol. 39, No. 2, pp. 143-146, ISSN 0018-926X

[7] Tayem, N. & Kwon, H. (2005) L-shape 2-dimensional Arrival Angle Estimation with Propagator Method. *IEEE Trans. Antennas and Propagation*, Vol. 53, No. 5, pp. 1622-1630, ISSN 0018-926X

[8] Harabi, F.; Gharsallah, A. & Marcos, S.(2009) Three-dimensional Antennas Array for the Estimation of Direction of Arrival, *IET Microwaves, Antennas and Propagation*. Vol. 3, No. 5, pp. 843-849, ISSN 1751-8725

[9] Liang, J. & Liu, D. (2010) Joint Elevation and Azimuth Direction Finding Using L-shaped Array. *IEEE Trans. Antennas and Propagation*, Vol. 58, No. 6, pp. 2136-2141, ISSN 0018-926X

[10] Pesavento, M.; Gershman, A. & Wong, K.(2002) Direction finding in partly calibrated sensor arrays composed of multiple subarrays, *IEEE Trans. Signal Processing*, Vol. 50, No. 2, pp. 329–338, ISSN 1053-587X

[11] Gao, F. & Gershman, A. (2005) A generalized ESPRIT approach to direction- of- arrival estimation, *IEEE Signal Processing Letters*. Vol. 12, pp. 254–257. ISSN 1070-9908

[12] Liang, J.; etc. (2011)L-shaped array-based elevation and azimuth direction finding in the presence of mutual coupling. *Signal Processing*, Vol. 91, No. 5, pp. 1319-1328, ISSN 0165-1684

[13] Grosicki, E.; Abed-Meraim, K. & Hua, Y. (2005)A weighted linear prediction method for near-field source localization, *IEEE Trans. Signal Processing*, Vol. 53, No. 10, pp. 3651-3660, ISSN 1053-587X

[14] Yuen, N. & Friedlander, B. (1998) Performance analysis of higher order ESPRIT for localization of near-field sources, *IEEE Trans. Signal Processing*, Vol. 46, No. 3, (March 1998), pp.709-719, ISSN 1053-587X

[15] Challa, R. & Shamsunder, N. (1995) High-order subspace based algorithms for passive localization of near-field sources, *in: The Twenty-Ninth Asilomar Conference on Signals, Systems & Computers*, Pacific Grove, California, USA, vol. 2, pp. 777-781

[16] Liang, J. & Liu, D. (2010) Passive localization of mixed near-field and far-field sources using two-stage MUSIC algorithm. *IEEE Trans. Signal Processing*, Vol. 58, No. 1, pp. 108-120, ISSN 1053-587X

[17] Tichavsky, P.; Wong, K. & Zoltowski, M. (2001) Near-field/far-field azimuth elevation angle estimation using a single vector-hydrophone. *IEEE Trans. Signal Processing,* Vol. 49, No. 11, pp. 2498-2510, ISSN 1053-587X

[18] Doclo, S. & Moonen, M. (2003) Design of far-field and near-field broadband beamformers using eigenfilters. *Signal Processing,* Vol.83, No.12, pp. 2641-2673, ISSN 0165-1684

[19] Mukai, R.; Sawada, H. & Araki, S. etc.(2006) Frequency-domain blind source separation of many speech signals using near-field and far-field models, *EURASIP Journal on Applied Signal Processing,* Vol. 2006, Article ID 83683, pp. 1-13, ISSN 1687-0433

[20] Dogan, M. & Mendel, J. (1995)Applications of cumulants to array processing-Part I: aperture extension and array calibration, *IEEE Trans. Signal Processing,* Vol. 43, No. 5, pp. 1200-1216, ISSN 1053-587X

[21] Tayem, N. & Kwon, H. (2006) Azimuth and elevation angle estimation with no failure and no eigen decomposition, *Signal Processing,* Vol. 86, No. 1, pp. 8-16, ISSN 0165-1684

[22] Liu, T. & Mendel, J. (1998) Azimuth and elevation direction finding using arbitrary array geometries, *IEEE Trans. Signal Processing,* Vol. 46, No. 7, pp. 2061-2065, ISSN 1053-587X

[23] Liang, J. (2009) Joint azimuth and elevation direction finding using cumulant. *IEEE Sensors Journal,* Vol. 9, No. 4, pp. 390-398, ISSN 1530-437X

2

Experimental Calibration for Electronic Beamforming with Sensor Arrays

Lara del Val[1], María I. Jiménez[1], Alberto Izquierdo[1], Juan J. Villacorta[1] and Mariano Raboso[2]
[1]*Universidad de Valladolid, Departamento de Teoría de la Señal y Comunicaciones e Ingeniería Telemática,*
[2]*Universidad Pontificia de Salamanca (Escuela Universitaria de Informática), Spain*

1. Introduction

A sensor array system with N channels is assumed to have the same characteristics for each channel, which are composed of a sensor and an amplification system. In beamforming applications, the gain and phase of each channel are key elements in the synthesis of the beampattern (van Veen & Buckley, 1988). On the other hand, the position of the sensors in the array and the orientation of the axis/plane where the array is placed are also important for an accurate calculation of the weight vector.

In real implementation, specific parameters of each channel show a large deviation from their relative values. These errors exist due to small tolerances in sensor specifications or in the components in the amplification system, or even due to deviations in the position of the array sensors. These deviations, in the relative gain and phase of each channel, can produce errors in the pointing direction, as well as an increment in the sidelobe levels. These errors deteriorate the system performance (Barton, 2005), (Godara, 2004), (Naidu, 2001).

Usually, each channel's behaviour is unknown with enough accuracy and it can vary in time. Therefore, it is necessary to implement auto-calibration, which allows to cancel or to compensate the differences between each channel of the system (Skolnik, 2001). There are many analytic models, which are based on arrays with a large number of sensors (Barton, 2005), (Godara, 2004), (Swindlehurst, 1996), (Quazi, 1982), that allow to establish a relation, in a statistical sense, between the phase and gain errors of each sensor and the deviations in the radiation/reception pattern of the beamformer (particularly on the pointing angle, the main beam width or the sidelobe level). However, the errors with arrays with a small number of sensors do not fulfil those previous analytic models. Therefore, it is necessary to carry out detailed analysis of the degradations that are caused by these errors.

2. System description

A detection and position measurement system has been designed. The hardware of the proposed system is formed by:

- A PC with a Pentium processor, with one Innovative Integration M6713 card, which includes a C6713 DSP. M6713 hosts an Omnibus SD16 module, which provides 16 channels of 18 bit, 48 kHz sigma-delta A/D and D/A converters.
- A transmission (Tx) uniform linear array (ULA) formed by 8 amplifiers, each one with 2 channels, and 15 tweeters.
- A reception (Rx) ULA formed by 2 multi-channel preamplifiers, each with 8 channels, and 15 studio microphones.

Fig. 2-1 shows the acoustic transmission and reception arrays.

Fig. 2.1. Transmission (below) and reception (above) arrays

As the system uses low cost components, the calibration data of these components are not available. Therefore, it is important to know the dispersion of the characteristics of these components. An anechoic chamber has been built to do tests. Inside this chamber, the Tx and Rx arrays are located at one of the sides of the chamber. A Tx sensor and a Rx sensor, which have been taken as reference, are located at the opposite side, at a distance of 4.2 m. The hardware implements a narrow band system that detects the existence and also the position of targets in different environments, on the basis of multi-function radars (Barton, 2005), (Sabatini & Tarantino, 1994), (Billeter, 1989). Electronic beams are implemented, using beamforming techniques (Naidu, 2001), (Van Veen & Buckley, 1988), from the in-phase and quadrature signals of each channel. Fig. 2-2 shows the employed signal processing algorithm.

The main blocks are described as follows:

- Transmitter: The Tx beamforming block generates a signal for a steering angle for each sensor. The signal frequency is 7 kHz. The Tx phase and gain compensation block compensates the gain and phase of each sensor and channel of the transmitter.
- Receiver: A band pass filter with a frequency range from 6 kHz to 8 kHz. The I+Q demodulator block obtains the in-phase and quadrature components of the signals. The Rx phase and gain compensation block compensates the gain and phase of each sensor and channel in the receiver. The Rx beamforming block processes the signals received for each sensor to form a beam at a specific steering angle. The matched filter block applies an optimal filter to maximize the signal-to-noise ratio (SNR).

This calibration method employs a reference sensor located opposite to the array on a perpendicular axis to the plane of the array (broadside). Two types of calibration are defined:

- Calibration of the Tx array, using a reference microphone
- Calibration of the Rx array, using a reference loudspeaker (tweeter).

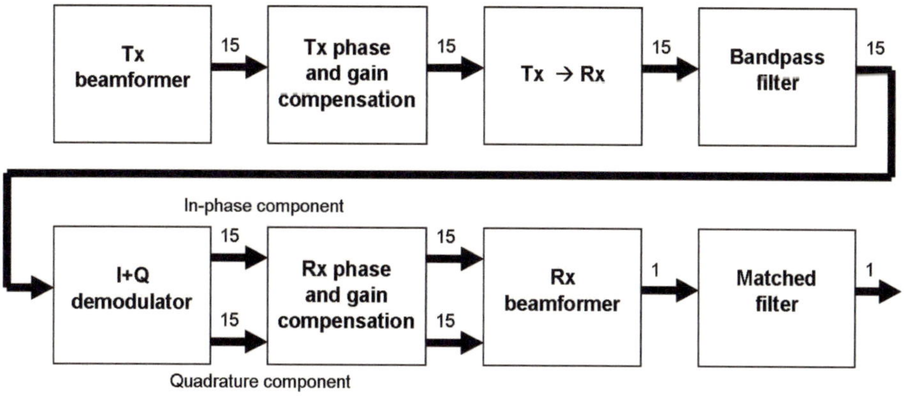

Fig. 2.2. Signal processing algorithm

The calibration system is based on the hypothesis of working with spherical waves, because the acoustic system is used under near field conditions. Since spherical waves are assumed, signals that are transmitted/received between the sensors of the array and the reference sensors arrive with different gains and phases. These gains and phases are taken into account in calibration. As the position of the reference sensor is known, the theoretical amplitudes and phases of the signal of each channel can be calculated, and then compared with real data, extracting the information to carry out the phase and gain compensation.

3. Transmitter array calibration

A sinusoidal pulse of 7 kHz and 5 ms width is used with a reference microphone located at 4.2 m range and 0° from the array perpendicular angle. Pulse signals are transmitted sequentially by each loudspeaker of the transmission array. Then, they are received in the reference microphone and, finally, are processed together. Experimental signals received at loudspeakers number 0, 3 and 13 are presented in Fig. 3-1.

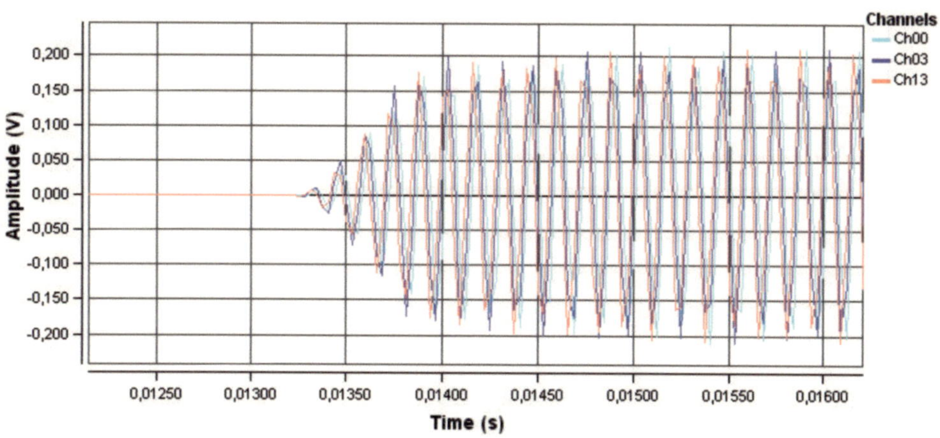

Fig. 3.1. Received signals from the Tx array

Module and phase values of the signals can be obtained using the described processing algorithm. Module values are showed in Fig. 3-2.

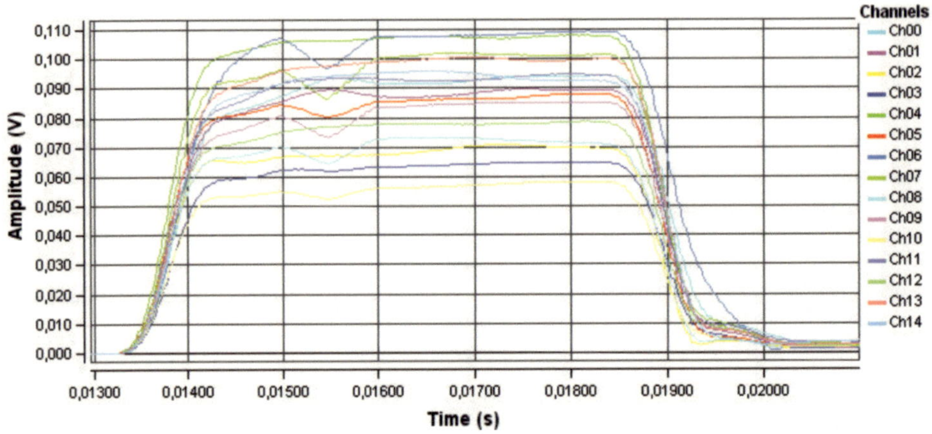

Fig. 3.2. Tx module values

The average of the module is calculated for the pulse length. Relative gains are obtained by normalising the module averages of each channel with respect to channel 0. They are showed in Table 3-1.

0	1	2	3	4	5	6	7
1.00	0.95	0.74	0.69	1.16	0.92	1.15	1.06

8	9	10	11	12	13	14
0.77	0.89	0.61	1.01	0.84	1.07	1.02

Table 3.1. Tx relative gains

In the same way, phase values of each sensor are obtained, using channel 0 as reference. They are showed in Fig. 3-3.

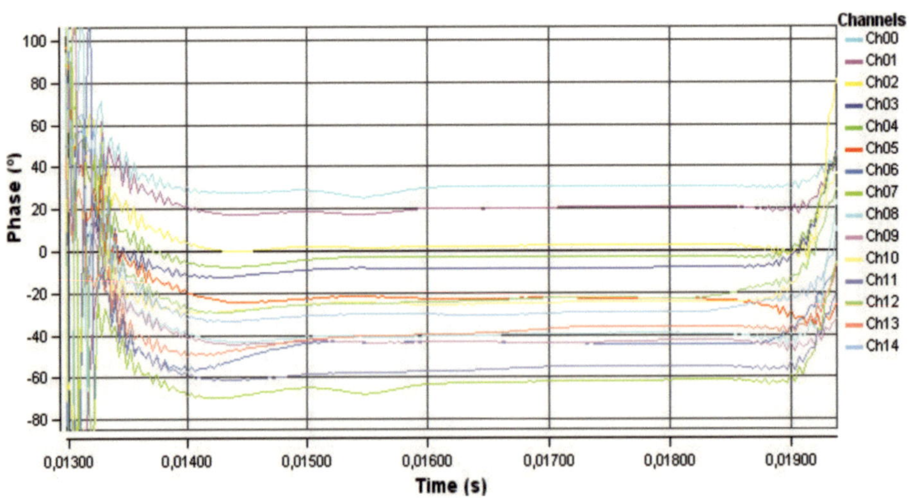

Fig. 3.3. Tx phase values

The average of the phase for each channel is calculated for the pulse length, and relative phases are obtained. They are shown in Table 3-2.

0	1	2	3	4	5	6	7
-49.0	-45.8	-52.3	-53.9	-41.5	-56.1	-74.2	-92.6

8	9	10	11	12	13	14
-70.5	-76.8	-63.3	-101.6	-78.0	-103.8	-108.4

Table 3.2. Tx relative phases (º)

The beampattern is calculated by steering the array from -90º to 90º, using classical beamforming techniques, and obtaining the maximum value from the matched filter output. Fig. 3-4 shows the obtained beampattern when the relative gains and phases of the sensors are not compensated. Fig. 3-5 shows the beampattern with relative gain and phase compensations before the beamforming is applied.

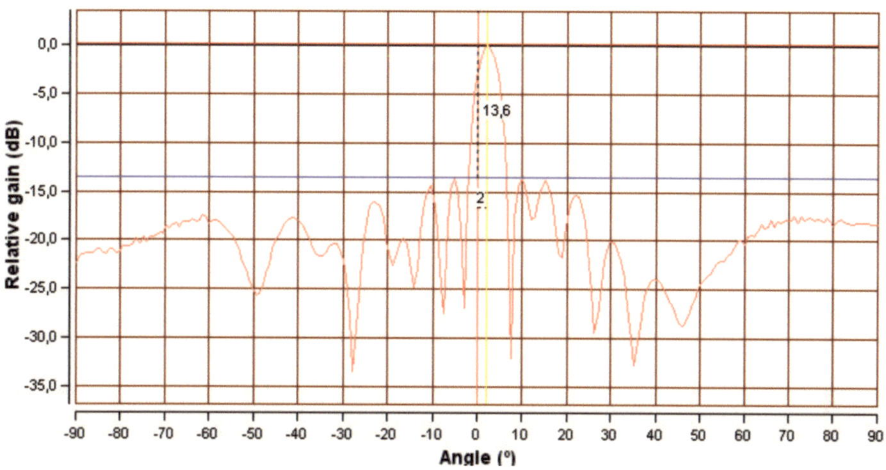

Fig. 3.4. Tx beampattern without compensation

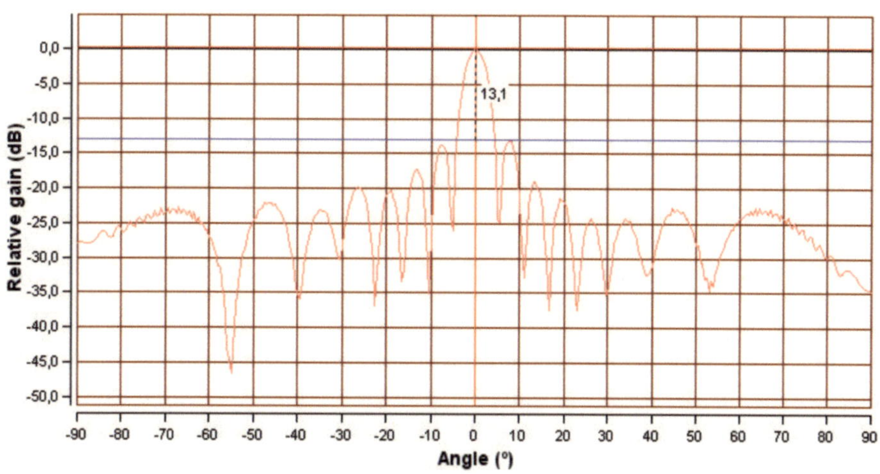

Fig. 3.5. Tx beampattern with compensation for 0º

Parameters obtained with and without gain and phase compensations are shown in Table 3-3.

	Without compensation	With compensation
Aiming error	2º	0º
Sidelobe level	13.6dB	13.1dB

Table 3.3. Tx beampattern parameters

To analyse if the calibrations of relative module and phase were independent of the steering angle, the system has been calibrated with the data obtained with the reference microphone placed at 0º, but with the reference microphone positioned now at -15º (with the same range). The obtained beampattern is presented in Fig. 3-6, where it is shown that the relative phase of the sensors of the Tx array under test does not depend on the steering angle. Therefore, the calibration in one angle is enough.

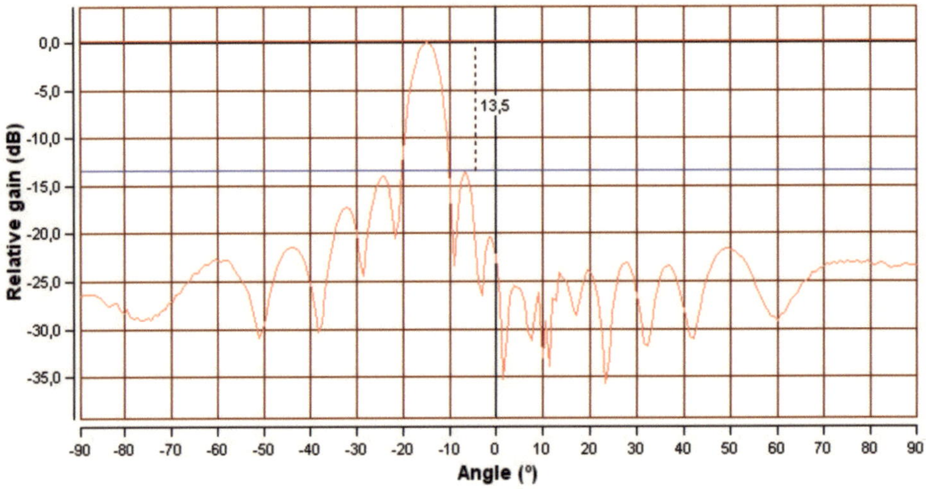

Fig. 3.6. Tx beampattern pointing to -15º with compensation for 0º

4. Receiver array calibration

A sinusoidal pulse of 7 kHz and 5 ms width is used with a reference loudspeaker located at 4.2 m range and 0º from the axis normal to the array. Also, in this case, spherical waves are assumed. A pulse signal is transmitted by the reference loudspeaker, and immediately, the signals received in the microphone array are recorded. The module and phase of the received signals can be obtained by means of the processing algorithm. The obtained module values are shown in Fig. 4-1.

Relative gains are obtained by means of the amplitude normalisation of each channel regarding channel 0. These relative gains are shown in Table 4-1. In the same way, phase values are obtained, using channel 0 as reference. These phases are shown in Fig. 4-2. The average of the phase of each channel is calculated for the pulse length. Relative phase values are obtained and shown in Table 4-2.

The beam-pattern is calculated by steering the array from -90º to 90º, using the classic beam-forming techniques, and obtaining the maximum value from the matched filter output. Fig. 4-3 shows the beam-pattern when the relative gain and phase of the sensors are not compensated. Fig. 4-4 shows the corresponding beampattern when relative gain and phase compensation is employed. The parameter values obtained with and without gain and phase compensation are shown in Table 4-3.

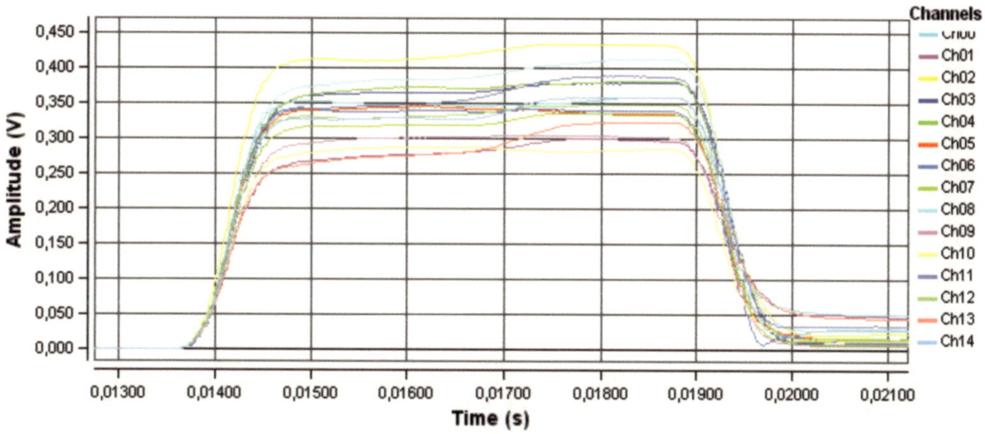

Fig. 4.1. Rx module values

0	1	2	3	4	5	6	7
1.00	0.82	1.21	1.06	1.07	0.97	0.97	0.93

8	9	10	11	12	13	14
1.13	0.86	0.81	1.05	0.97	0.84	0.97

Table 4.1. Rx relative gains

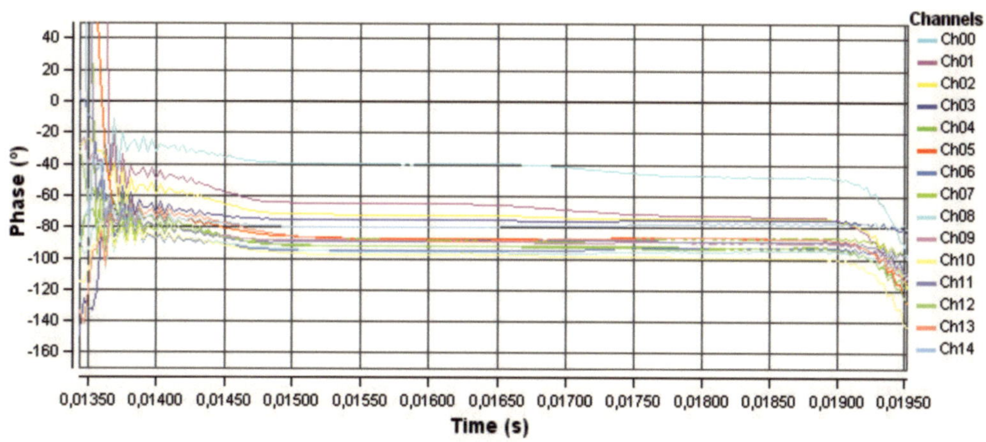

Fig. 4.2. Rx phase values

0	1	2	3	4	5	6	7
-29.6	-47.5	-45.6	-42.3	-54.8	-46.7	-52.1	-45.1
8	9	10	11	12	13	14	
-53.7	-49.0	-60.9	-55.6	-65.2	-66.7	-65.6	

Table 4.2. Rx relative phases (°)

	Without compensation	With compensation
Aiming error	2°	0°
Sidelobe level	12.6dB	13.2dB

Table 4.3. Rx beampattern parameters

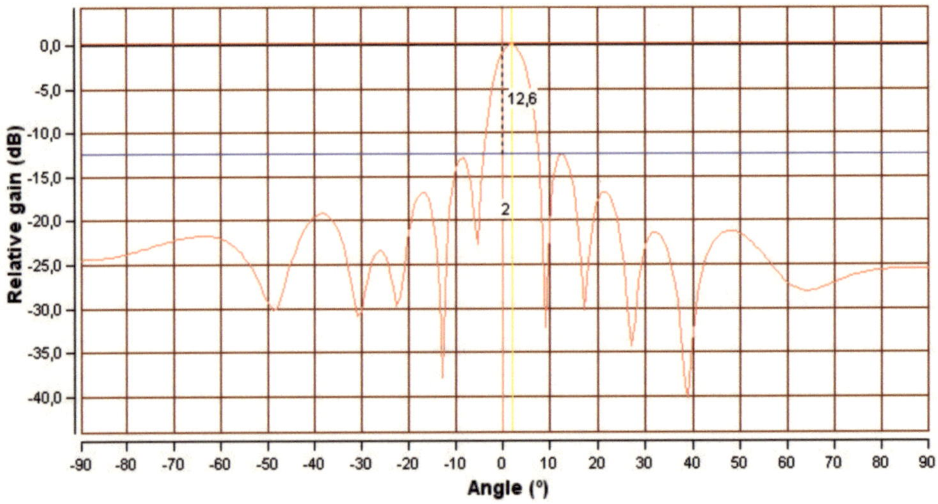

Fig. 4.3. Rx beampattern without compensation

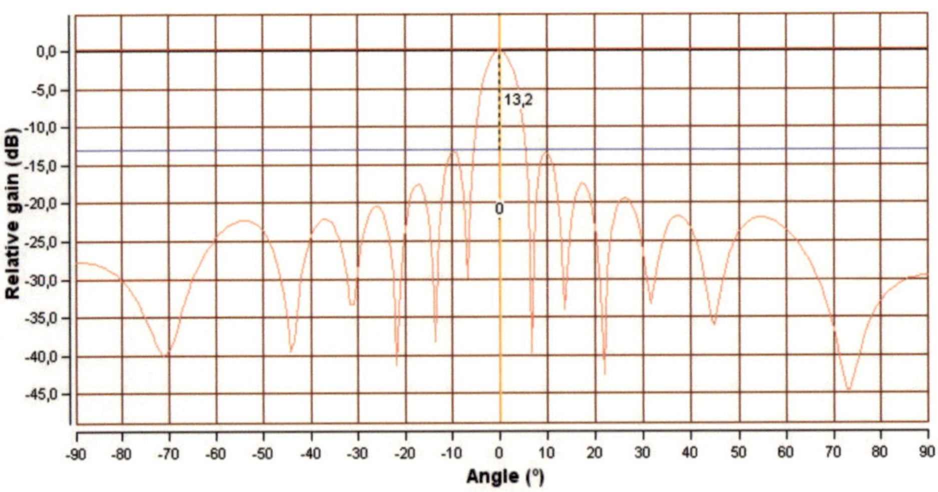

Fig. 4.4. Rx beampattern with compensation for 0°

Using the calibration data obtained with the loudspeaker pointing to with the reference loudspeaker placed now at -15° (in the same range), the independence of module and phase calibrations from the steering angle is analysed. The beampattern obtained is shown in Fig. 4-5. It shows that relative phases of the Rx sensors do not depend on the steering angle. Therefore, calibrating in one angle is enough.

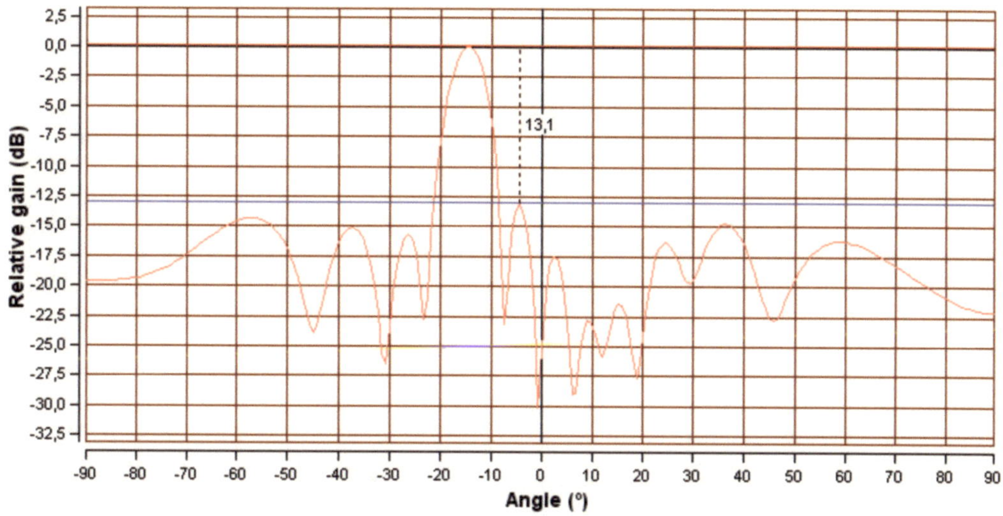

Fig. 4.5. Rx beampattern receiving at -15° with 0° compensation

5. Verification of calibration independence from steering angle

To check the independence of the phase and gain calibrations from the steering angle, these calibration methods have been proved in the same radar system, but with different transmitter and receiver arrays. In this case, the acoustic transmitter array is composed of 8 tweeters, while the acoustic reception array is composed of 8 electret condenser microphones. Fig. 5-1 shows the acoustic Tx and Rx arrays employed for verification of the calibration independence from the steering angle.

Fig. 5.1. Transmission (above) and reception (below)

5.1 Transmitter array calibration

Actually, the applied calibration method is exactly the same as shown in Section 3. It applies the same reference microphone, located opposite to the Rx array on a perpendicular axis to its plane (broadside). The only difference is that in this case, a sinusoidal pulse of 6 kHz and 0.5 ms width is used, and the band-pass filter employed has a frequency range from 5 kHz to 7 kHz. Normalising the amplitude of each channel, regarding to channel 0, the relative gains are obtained and presented on Table 5-1.

0	1	2	3	4	5	6	7
1.00	0.58	0.68	0.77	0.82	0.79	0.45	1.02

Table 5.1. Tx relative gains

The average of the phase for each channel is calculated over the pulse length, and the following relative phase table is obtained.

0	1	2	3	4	5	6	7
0.00	33.3	12.2	15.	50.6	44.2	34.8	55.9

Table 5.2. Tx relative phase

The beampattern is calculated by steering the array from -90° to 90°. Fig. 5-2 shows the beampattern when the relative gain and phase of the sensors are not compensated. Fig. 5-3 shows the corresponding beampattern when relative gain and phase compensation is employed. The parameter values obtained with and without gain and phase compensations are shown in Table 5-3.

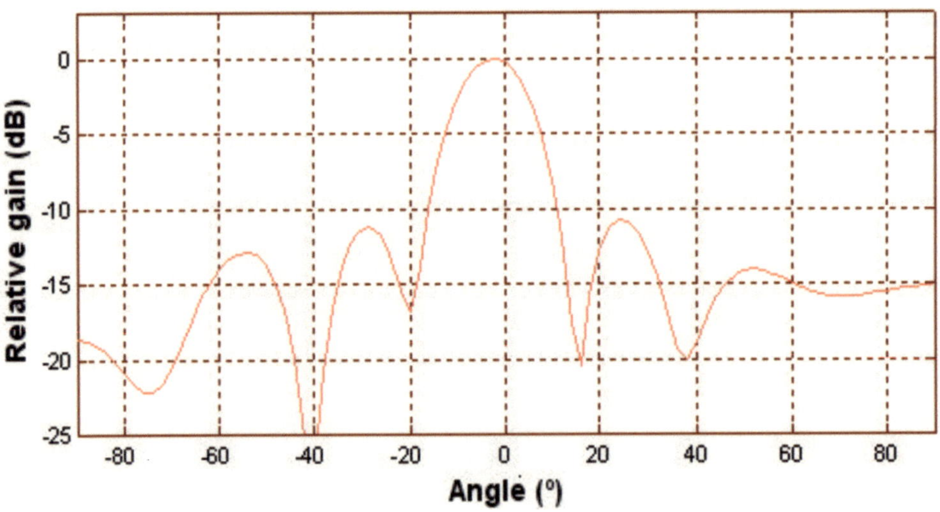

Fig. 5.2. Tx 0° beampattern without compensation

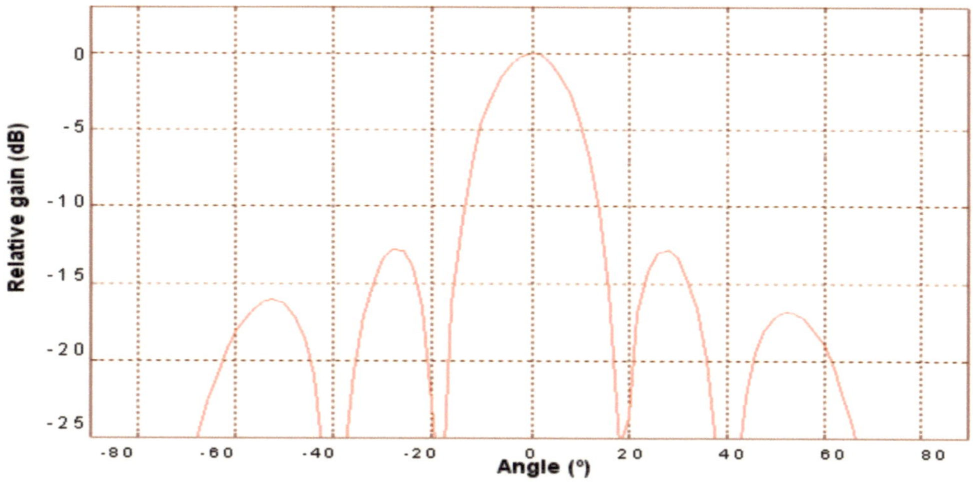

Fig. 5.3. Tx 0 ° beampattern with compensation

	Without compensation	With compensation
Aiming error	2°	0°
Sidelobe level	10.8dB	12.75dB

Table 5.3. Tx beampattern parameters

Using the calibration with a 0° reference microphone and the reference microphone placed now at -15° (in the same range), the beampattern obtained is presented in Fig. 5-4.

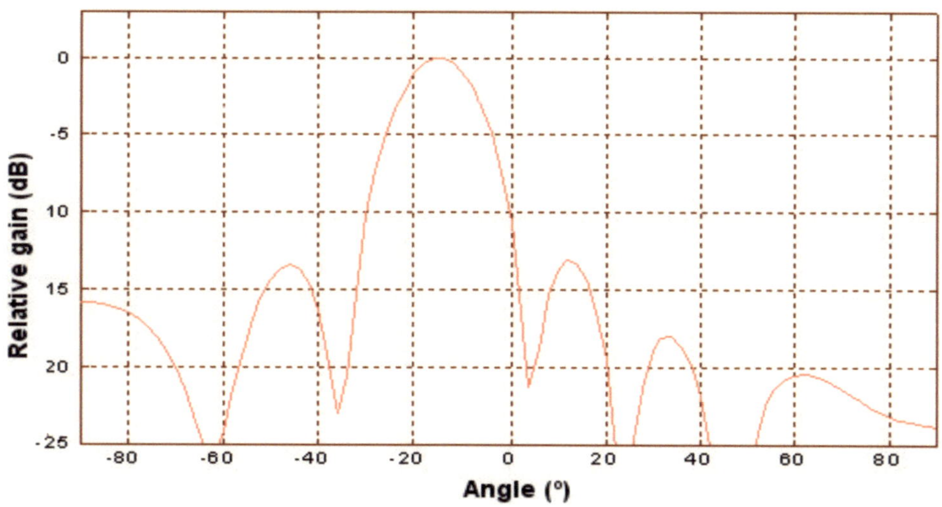

Fig. 5.4. Tx (-15) ° beampattern with 0° compensation

As shown in Section 4, the relative phases of the sensors of the Tx array do not depend on the steering angle. Therefore, calibrating the Tx array for a single angle is enough.

5.2 Receiver array calibration

In this case, a sinusoidal pulse of 6 kHz and 0.5 ms width is used with a reference loudspeaker. Normalising the amplitude of each channel, regarding to channel 0, relative gains are obtained, which are presented on Table 5-4.

0	1	2	3	4	5	6	7
1.00	0.86	0.59	0.98	0.92	0.65	1.12	1.09

Table 5.4. Rx relative gains

The average of the phase for each channel is calculated over pulse length, and the following relative phase table is obtained.

0	1	2	3	4	5	6	7
0.00°	33.3°	12.2°	15.0°	50.6°	44.2°	34.8°	55.9°

Table 5.5. Rx relative phases

Also in this case, the beampattern is calculated by steering the array from -90° to 90°, using classical beamforming techniques, and obtaining the maximum value from the matched filter output. Fig. 5-5 shows the beampattern when the relative gain and phase of the sensors are not compensated. Fig. 5-6 shows the corresponding beampattern when relative gain and phase compensation is employed. The parameter values that have been obtained with and without gain and phase compensations are shown in Table 5-6.

	Without compensation	With compensation
Aiming error	-2°	0°
Sidelobe level	7.8dB	12.75dB

Table 5.6. Rx beampattern parameters

As the aim of this section is to analyse the calibration independence from the steering angle, the system has been calibrated with the data obtained with the reference tweeter placed at 0°, but with the reference tweeter placed at -15° (with the same range). The obtained beam-pattern is showed in Fig. 5-7. It shows that the relative phases of the microphones that compose the Rx array under test depend on the steering angle. For this Rx array, the behaviour of the relative phases of its microphones is different. In this case, it is very important to calibrate the Rx array for each steering angle.

To solve this problem, a new module and phase calibration has been made for pre-compensation of the theoretical phase for a -15° angle, and obtaining relative phase averages and gains, which are showed on Table 5-7. With these new values, the beampattern is recalculated and showed in Fig. 5-8. Therefore, in this case, for the Rx array, it is necessary to calibrate each steering angle used.

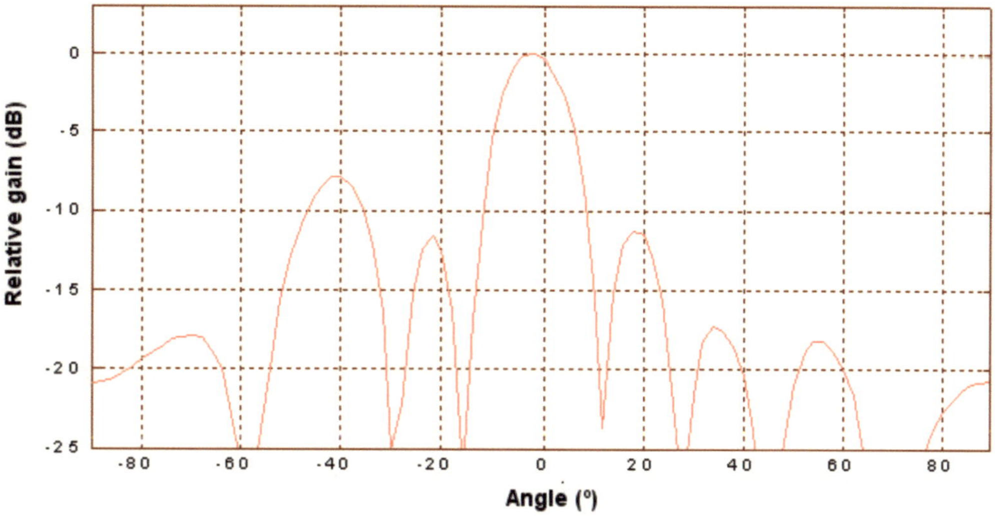

Fig. 5.5. Rx beampattern without compensation

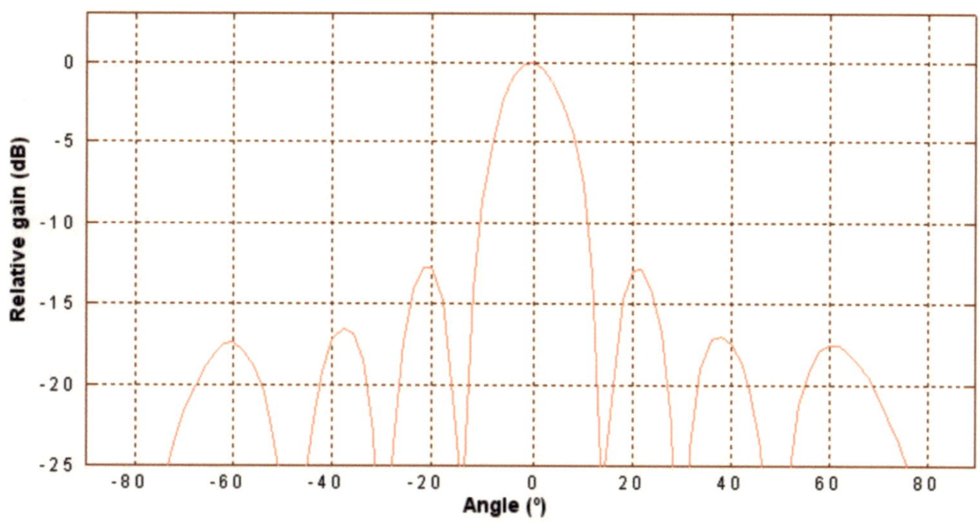

Fig. 5.6. Rx 0 ° beampattern with compensation for 0°

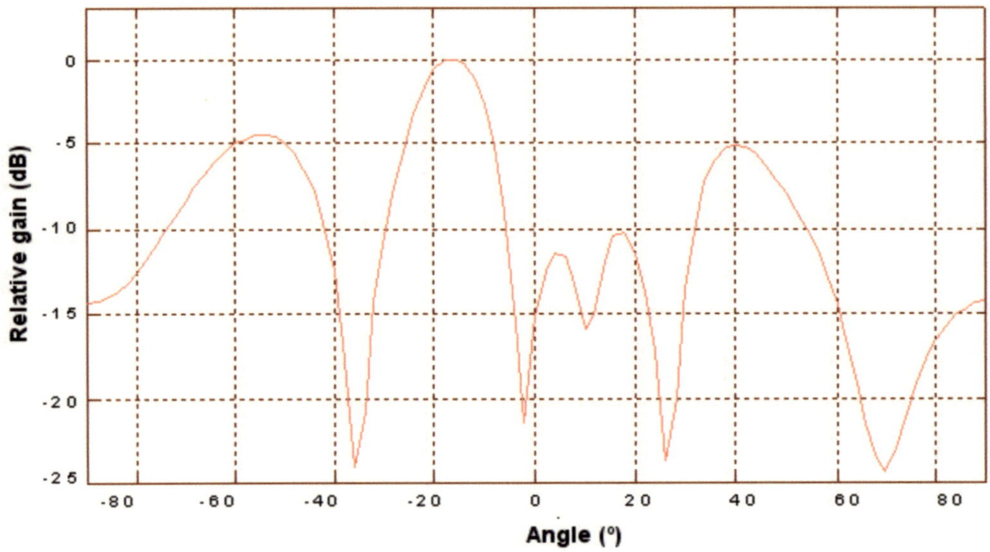

Fig. 5.7. Rx (-15°) beampattern with compensation for 0°

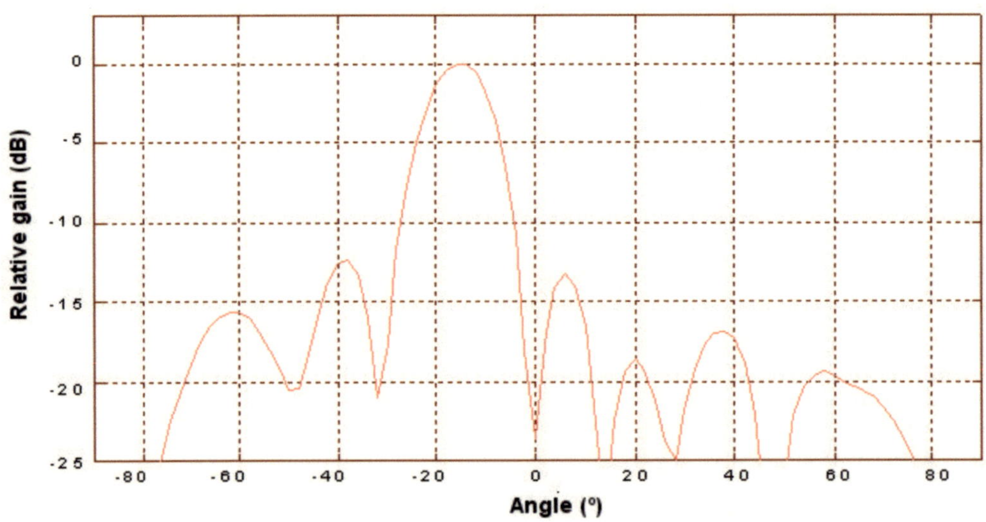

Fig. 5.8. Rx (-15) ° beampattern with compensation for (-15) °

	0	1	2	3	4	5	6	7
Amplitudes	1.00	0.86	0.59	0.98	0.92	0.65	1.12	1.09
Phases	0.00°	42°	87°	-8°	-277°	-267°	-42°	-312°

Table 5.7. Relative phases and gains

6. Conclusions

In this chapter, a method to calibrate sensor arrays to be employed as part of narrow band acoustic radars is shown. The proposed method allows obtaining compensation values, for both gain and phase, and for each sensor. Therefore, the obtained beampatterns improve considerably regarding to those without calibration. Beampatterns obtained with calibration approach the theoretical ones.

This work has shown that the independence of the relative phases of the sensors of the transmission and reception arrays from the steering angle depends on the particular arrays that are employed in the radar system. In other cases, the relative phases do not vary with the steering angle. In this case, a calibration in a unique angle is enough to obtain a good array performance; in some other cases, the calibration values depend on the steering angle, and a calibration for each steering angle used must be done.

7. References

Barton, D. (2005) *Radar System Analysis and Modelling*, Artech House, ISBN 9781580536813, Norwood, MA.

Billeter, D. (1989) *Multifunction array radar*, Artech House, ISBN 978-0890063590, Norwood, MA.

Godara, L. (2004) *Smart Antennas*, ser. The Electrical Engineering and Applied Signal Processing, CRC Press LLC, ISBN 9780849312069, Boca Ratón, Florida, USA.

Naidu, P. (2001) *Sensor Array Signal Processing*, ser. The Electrical Engineering and Applied Signal Processing. CRC Press LLC, ISBN 9780849311956, Boca Ratón, Florida, USA.

Quazi, A. (1982) Array beam response in the presence of amplitude and phase fluctuations, *Journal of the Acoustic Society of America*, vol. 72, pp. 171-180, ISSN 0030-3941.

Sabatini, S., & Tarantino, M. (1994) *Multifunction array radar: System design and analysis*, Artech House, ISBN 9780890065761, Norwood, MA.

Skolnik, M. (2001) *Introduction to Radar Systems* (3rd Ed.), ser. Electrical and Engineering. McGraw Hill International Editions, ISBN 9780072909807, New York, USA.

Swindlehurst, A. (1996) A Maximum a Posteriori Approach to Beamforming in the Presence of Calibration Errors, *in Proceedings of the 8th IEEE Signal Processing Workshop on Statistical Signal and Array Processing*, pp. 82-85 ISBN 0-8186-7576-4, Corfú, Grecia, June 1996.

van Veen, B., & Buckley, K. (1988) Beamforming: a versatile approach to spatial filtering, *IEEE ASSP Magazine*, vol.5, no.2, pp.4-24, ISSN 0740-7467.

3

Predictive Angle Tracking Algorithm Based on Extended Kalman Filter

Sheng-Yun Hou[1], Shun-Hsyung Chang[2] and Hsien-Sen Hung[3]
[1]*Department of Electrical Engineering, Hwa Hsia Institute of Technology*
[2]*Department of Microelectronics Engineering, National Kaohsiung Marine University*
[3]*Department of Electrical Engineering, National Taiwan Ocean University*
Taiwan

1. Introduction

In the case of moving sources, various target angle tracking algorithms have been proposed and reported in the literature for multiple narrow-band targets. Yang and Kaveh proposed an iterative adaptive eigen-subspace method in conjunction with the multiple signal classification (MUSIC) algorithm to track the DOA angles of multiple targets (Yang & Kaveh, 1988). Due to the data association problem caused by multi-target tracking, the adaptive MUSIC method fails to track targets when they are moving close to each other. Although the method proposed by Sword, et al. (1990) can avoid the data association problem, errors are accumulated in each iteration, making it unable to track targets that are mutually close. Due to the nature of prediction-correction filtering process, Kalman filter (KF) can reduce estimation errors and avoid the data association problem when applied to angle tracking, as stated in several references (Javier & Sylvie, 1999; Yang, 1995; Park, et al. 1994). Rao, et al. (1994) proposed to estimate DOA angles of targets using the maximum likelihood method and feeding the results to a KF. However, it is assumed that the signal powers of the targets are all different, making the algorithm impractical. Javier and Sylvie (1999) suggested to estimate target angles using the projection approximation subspace tracking algorithm with deflation (PASTd) (Yang, 1995) and a Newton-type method (for MUSIC spectrum) for the use in the KF. It has lower computational load and better tracking performance than Rao's algorithm, but still exhibits poor tracking success rate at low signal-to-noise ratios (SNRs). Park, et al. (1994) proposed an approach, which utilizes predicted angles obtained from Sword's method. The approach also uses the constrained least-squares criterion to confine the dynamic range of angles. The choice of relevant parameters is empirical and is not suitable for various scenarios of different moving speeds and SNRs. Besides, the tracking performance degrades seriously with an increasing number of crossing targets. Later on, to improve Park's method, Ryu, et al. (1999, 2002) suggested to obtain the angle innovations of the targets from a signal subspace, instead of the sensor output covariance matrix, via projection approximation subspace tracking (PAST) algorithm (Yang, 1995). Chang, et al. (2005) modified Park's algorithm by incorporating a spatial smoothing (Shan et al., 1985) technique to overcome multipath interference, and also coherent signal-subspace (CSS) (Wang & Kaveh, 1985) processing for tracking wideband targets. All of the above algorithms are based on the sample covariance matrix or signal subspace made with multiple snapshots of data from a sensor array. However, they all fail to track multiple targets when only a single snapshot measurement is available between two consecutive time steps during the tracking process, because DOA estimation using subspace-based approach requires sample covariance matrix or signal subspace with a rank of more than one.

In the case of a single snapshot measurement within each time increment, tracking multiple targets becomes feasible if the sensor array output is directly used as the measurement data in the extended Kalman filter (EKF) (Kong & Chun, 2000). However, the EKF is an approximate nonlinear state estimation technique with first-order linearization accuracy, and is suitable for the tracking problem since the measurement model is nonlinear in terms of the angles (states) to be estimated. The algorithm proposed by Kong and Chun (2000) exhibits low tracking success rate when targets approach near the points of intersection. The reason for this weakness is the EKF can be difficult to tune and often gives unreliable estimates if the system nonlinearities are severe.

2. Tracking algorithm

For tracking non-stationary targets efficiently and effectively, the predictive angle tracking algorithm based on extended Kalman filter (PAT-EKF) is presented. In the proposed algorithm, the sensor array output is used as measurement data in EKF, since the measurement model is nonlinear in terms of angle estimates. Using the predicted angles, the PAT-EKF algorithm modifies Park's method to obtain angular innovation, from which the angle estimates are updated (smoothed) via Kalman gain.

2.1 Data model

In the data model, M targets moving (for tracking) in a plane are considered, which contain an array of L sensors (or hydrophones). The sensor positions are assumed to be known, and it takes them to be placed uniformly on a line with spacing of d between two adjacent sensors (abbreviated as ULA), measured in the unit of wavelength λ. The motion of the targets is assumed to be at constant angular speed in the presence of Gaussian disturbance, and is observed every T seconds. Let $\theta_m(t) \in [-\pi/2, \pi/2]$, measured clockwise with respect to y axis, denote the DOA angle of the mth target at time t. Assuming that these targets are located in the far field and their radiated signals are narrowband with a common angular frequency ω_0, the output of the lth isotropic sensor at time t is then

$$r_l(t) = \sum_{m=1}^{M} e^{-j\omega_o \tau_{lm}} s_m(t) + n_l(t) \tag{1}$$

where $s_m(t) \in R$ is the signal transmitted by the mth target at time t, $n_l(t)$ is a complex Gaussian white noise with zero mean and variance σ_n^2, which is uncorrelated with the target signals, and τ_{lm} is the difference in time delays of the mth target reaching the first (reference) sensor and the lth sensor.

By using vector-matrix representation, the output of the sensor array is given by

$$\mathbf{r}(t) = \mathbf{A}[\boldsymbol{\theta}(t)]\mathbf{s}(t) + \mathbf{n}(t) \tag{2}$$

where $\mathbf{r}(t)=[r_1(t),...,r_L(t)]^T$, $\mathbf{s}(t)=[s_1(t),...,s_M(t)]^T$, $\mathbf{n}(t)=[n_1(t),...,n_L(t)]^T$ are the output data, target signal, and noise vectors, respectively. $\boldsymbol{\theta}(t)=[\theta_1(t),\theta_2(t),...,\theta_M(t)]^T$ is the target DOA vector and $\mathbf{A}[\boldsymbol{\theta}(t)]$ is the array direction matrix with the direction vector of the mth target (the mth column vector)

$$\mathbf{a}_m=[1,e^{-j\frac{2\pi}{\lambda}d\sin\theta_m},...,e^{-j\frac{2\pi}{\lambda}(L-1)d\sin\theta_m}]^T, \quad m=1,\ldots,M. \tag{3}$$

Suppose that there are K measurements (snapshots) that are taken for each increment T, and the time increment is sufficiently small allowing us to approximate the target as stationary. Figure 1 shows the sensor array and source configurations in 2-D space.

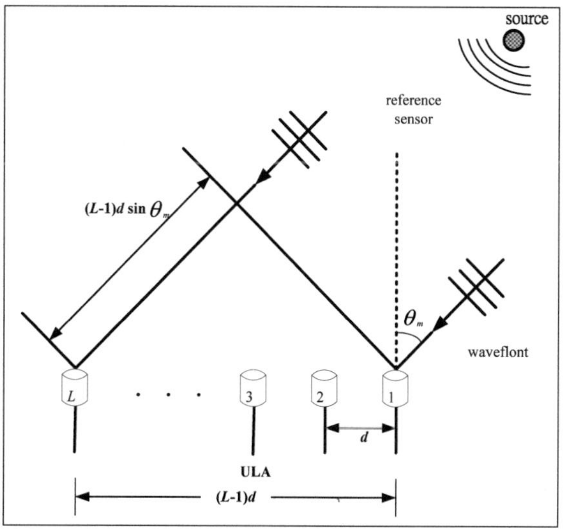

Fig. 1. Sensor array and source configurations in 2-D space

2.2 PAT-EKF algorithm

First, it depicts the discrete-time state (process) model for the target motion. For each time index k, it defines the state vector for the mth target as $\mathbf{x}_m(k)=\begin{bmatrix}\theta_m(k) & \dot{\theta}_m(k)\end{bmatrix}^T$, consisting of its DOA and angular speed. The mth target motion can lead to the process equation (Park et al., 1994)

$$\mathbf{x}_m(k+1)=\mathbf{F}\mathbf{x}_m(k)+\mathbf{w}_m(k) \tag{4}$$

$$\mathbf{F}=\begin{bmatrix}1 & T\\ 0 & 1\end{bmatrix}$$

where the process noise vector $\mathbf{w}_m(k)$ is assumed to be Gaussian distributed with zero mean and covariance

$$\mathbf{Q}_m=\sigma_w^2\begin{bmatrix}\frac{T^3}{3} & \frac{T^2}{2}\\ \frac{T^2}{2} & T\end{bmatrix}$$

Assume that the motion of each target is mutually independent. By defining the composite state vector as $\mathbf{x}(k)=\left[\mathbf{x}_1^T(k),...,\mathbf{x}_M^T(k)\right]^T$, the system dynamics is governed by the process model

$$\mathbf{x}(k+1)=\bar{\mathbf{F}}\mathbf{x}(k)+\mathbf{w}(k) \tag{5}$$

$$\bar{\mathbf{F}}=diag(\overbrace{\mathbf{F},\cdots,\mathbf{F}}^{M})$$

The process noise vector $\mathbf{w}(k)$=$\left[\mathbf{w}_1^T(k),...,\mathbf{w}_M^T(k)\right]^T$ reflects the random modeling error, which is Gaussian distributed with zero mean vector and covariance

$$\bar{\mathbf{Q}}=diag(\mathbf{Q}_1,\cdots,\mathbf{Q}_M)$$

The matrices $\bar{\mathbf{F}}$ and $\bar{\mathbf{Q}}$ are all block diagonal. Although the process equation is a linear model, the measurement model of (2) is a vector nonlinear function of the target DOAs (and thus, of the target state vectors as well), which can be restated as

$$\mathbf{r}(k)\triangleq\mathbf{h}(\mathbf{x}(k),\mathbf{s}(k),\mathbf{n}(k))=\mathbf{A}(\mathbf{x}(k))\mathbf{s}(k)+\mathbf{n}(k) \tag{6}$$

where $\mathbf{n}(k)$ is complex Gaussian noise process with the known covariance $\sigma_n^2\mathbf{I}$, and is assumed to be uncorrelated with the process noise $\mathbf{w}(k)$. Assuming that a uniform linear array of L sensors with a half wavelength of inter-element spacing d is deployed, the partial derivative (Jacobian) matrix of the measurement model (6) is given by

$$\mathbf{H}(k)=\frac{\partial\mathbf{h}}{\partial\mathbf{x}}=[\mathbf{H}_1(k),...,\mathbf{H}_M(k)]$$

By augmenting the real and imaginary parts of each complex matrix $\mathbf{H}_m(k)$, it has the composite real matrix of dimension $2L\times 2M$

$$\bar{\mathbf{H}}(k)=\begin{bmatrix} real(\mathbf{H}_1(k),...,\mathbf{H}_M(k)) \\ imag(\mathbf{H}_1(k),...,\mathbf{H}_M(k)) \end{bmatrix}$$

which can be expressed as

$$\bar{\mathbf{H}}(k)=\begin{bmatrix} 0 & 0 & \cdots & 0 & 0 \\ g_{1,1} & 0 & \cdots & g_{M,1} & 0 \\ \vdots & \vdots & \ddots & \vdots & \vdots \\ g_{1,L-1} & 0 & \cdots & g_{M,L-1} & 0 \\ 0 & 0 & \cdots & 0 & 0 \\ c_{1,1} & 0 & \cdots & c_{M,1} & 0 \\ \vdots & \vdots & \ddots & \vdots & \vdots \\ c_{1,L-1} & 0 & \cdots & c_{M,L-1} & 0 \end{bmatrix}$$

where

$$g_{m,b} = -\sin(\pi b \sin(\theta_m(k)))\cos(\theta_m(k))s_m(k) \tag{7}$$

$$c_{m,b} = -\cos(\pi b \sin(\theta_m(k)))\cos(\theta_m(k))s_m(k) \tag{8}$$

m=1, …, M, b=1, …, L−1.

Initially (at k=0), the target DOA angles, $\{\hat{\theta}_m(-1)\}$ and $\{\hat{\theta}_m(0)\}$ at two successive time instants, k=−1 and k=0, are assumed to be available, which can be estimated by any subspace-based DOA angle estimation algorithm for instance the MUSIC algorithm (Schmidt, 1986). Thus, the initial state vector estimate can be set as $\hat{\mathbf{x}}(0|0) = [\hat{\theta}_1(0)\ \ (\hat{\theta}_1(0) - \hat{\theta}_1(-1))/T, \cdots,\ \ \hat{\theta}_M(0)\ \ (\hat{\theta}_M(0) - \hat{\theta}_M(-1))/T]^T$ with its covariance matrix $\mathbf{P}(0|0)$, given by

$$\mathbf{P}(0|0) = \sigma_v^2 \begin{bmatrix} 1 & \frac{1}{T} & & & \mathbf{0} \\ \frac{1}{T} & \frac{2}{T^2} & & & \\ & & \ddots & & \\ & & & \ddots & \\ & & & 1 & \frac{1}{T} \\ \mathbf{0} & & & \frac{1}{T} & \frac{2}{T^2} \end{bmatrix}$$

For k=1,2,…, the proposed tracking algorithm can be summarized in the following four steps.

Step 1. Prediction of angles

The prediction of the state vector and its covariance matrix can be obtained from the existing estimates by the equations

$$\hat{\mathbf{x}}(k|k-1) = \overline{\mathbf{F}}\hat{\mathbf{x}}(k-1|k-1) \tag{9}$$

$$\mathbf{P}(k|k-1) = \overline{\mathbf{F}}\mathbf{P}(k-1|k-1)\overline{\mathbf{F}}^T + \overline{\mathbf{Q}} \tag{10}$$

The first element of each state vector $\hat{\mathbf{x}}_m(k|k-1)$ is the predicted estimate $\hat{\theta}_m(k|k-1)$ of $\theta_m(k)$. The predicted direction matrix $\mathbf{A}(k|k\text{-}1)$ can be obtained by (3) using $\hat{\theta}_m(k|k-1)$ for $\theta_m(k)$. From (2), the predicted output of the sensor array becomes

$$\mathbf{r}(k|k-1) = \mathbf{A}(k|k-1)\mathbf{s}(k) \tag{11}$$

and can be obtained once $\mathbf{s}(k)=[s_1(k),\ldots, s_M(k)]^T$ is estimated by invoking the maximum likelihood method as

$$\hat{\mathbf{s}}(k) = \left[\mathbf{A}^H(k|k-1)\mathbf{A}(k|k-1)\right]^{-1}\mathbf{A}^H(k|k-1)\mathbf{r}(k) \tag{12}$$

Step 2. Computations of the angle innovation

After time T, a new array output is observed and the direction matrix can be written as

$$\mathbf{A}(k) = \mathbf{A}(k \mid k-1) + \delta\mathbf{A}(k) \tag{13}$$

where $\delta\mathbf{A}(k)$ is the error matrix, which can be derived, according to (Sword *et al.*, 1990), as

$$\delta\mathbf{A}(k) = \begin{bmatrix} 0 & \cdots & 0 \\ \delta\gamma_1 & \cdots & \delta\gamma_M \\ 2\gamma_1\delta\gamma_1 & \cdots & 2\gamma_M\delta\gamma_M \\ \vdots & \ddots & \vdots \\ (L-1)\gamma_1^{L-2}\delta\gamma_1 & \cdots & (L-1)\gamma_M^{L-2}\delta\gamma_M \end{bmatrix} \tag{14}$$

wherein

$$\delta\gamma_m(k) = -j\pi\cos\theta_m(k)\gamma_m(k)\delta\theta_m(k) \tag{15}$$

Thus, the residual array output $\Delta\mathbf{r}(k)$ can be obtained and written as

$$\Delta\mathbf{r}(k) = \mathbf{r}(k) - \mathbf{r}(k \mid k-1) = \delta\mathbf{A}(k)\mathbf{s}(k) + \mathbf{n}(k) \tag{16}$$

Note that the first row vector of $\delta\mathbf{A}(k)$ in (14) is a null vector. To reduce the computation, the null vector allows us to define a $(L-1)\times 1$ vector $\Delta\tilde{\mathbf{r}}$, which is obtained by removing the first element of $\Delta\mathbf{r}$ in (16). By substituting (15) into (14), $\Delta\tilde{\mathbf{r}}$ can be represented by (dropping k temporarily)

$$\Delta\tilde{\mathbf{r}} = \mathbf{B}\delta\boldsymbol{\theta} + \tilde{\mathbf{n}} \tag{17}$$

In (17), the $(L-1)\times M$ matrix $\mathbf{B}$ is

$$\mathbf{B} = -j\pi \begin{bmatrix} \cos(\theta_1)\gamma_1 s_1 & \cdots & \cos(\theta_M)\gamma_M s_M \\ 2\cos(\theta_1)\gamma_1^2 s_1 & \cdots & 2\cos(\theta_M)\gamma_M^2 s_M \\ \vdots & \ddots & \vdots \\ (L-1)\cos(\theta_1)\gamma_1^{L-1} s_1 & \cdots & (L-1)\cos(\theta_M)\gamma_M^{L-1} s_M \end{bmatrix}$$

where θ_m is substituted with the predicted angle $\hat{\theta}_m(k \mid k-1)$, and $\delta\boldsymbol{\theta} = [\delta\theta_1(k), \delta\theta_2(k), \cdots, \delta\theta_M(k)]$ is the unknown angle innovation vector to be estimated. In general, a least-squares solution of (17) is given by $\delta\boldsymbol{\theta} = (\mathbf{B}^H\mathbf{B})^{-1}\mathbf{B}^H\Delta\tilde{\mathbf{r}}$. However, the modified solution

$$\delta\boldsymbol{\theta} = (\mathbf{B}^H\mathbf{B} + \mathbf{L})^{-1}\mathbf{B}^H\Delta\tilde{\mathbf{r}} \tag{18}$$

as suggested in Park's algorithm, will be used to constrain the absolute values of innovations for the cases of nearby targets, where $\mathbf{L}$ is a weighting matrix with diagonal form.

Step 3. Estimation of the angles

The estimated angle can be obtained as

$$\hat{\theta}_m(k) = \hat{\theta}_m(k \mid k-1) + \delta\theta_m(k) \tag{19}$$

Furthermore, $\hat{\theta}_m(k)$ and $\hat{s}_m(k)$ are substituted into (7) and (8) to update the matrix $\bar{\mathbf{H}}(k)$.

Step 4. Smoothing the estimated angles

Since the state vector is real-valued, it formulates the state estimation equation as

$$\hat{\mathbf{x}}(k\,|\,k) = \hat{\mathbf{x}}(k\,|\,k-1) + \mathbf{K}(k)\Delta\bar{\mathbf{r}}(k) \tag{20}$$

where $\Delta\bar{\mathbf{r}}(k) = [\Delta\mathbf{r}_R\ \Delta\mathbf{r}_I]^T$ is a real vector; $\Delta\mathbf{r}_R$ and $\Delta\mathbf{r}_I$ are the real and imaginary parts of $\Delta\mathbf{r}(k)$ from (16). **K(*k*)** is the Kalman Gain matrix, given by

$$\mathbf{K}(k) = \mathbf{P}(k\,|\,k-1)\bar{\mathbf{H}}^T(k)\left[\bar{\mathbf{H}}(k)\mathbf{P}(k\,|\,k-1)\bar{\mathbf{H}}^T(k) + \sigma_n^2\mathbf{I}\right]^{-1} \tag{21}$$

The covariance matrix of $\hat{\mathbf{x}}(k\,|\,k)$ is given by

$$\mathbf{P}(k\,|\,k) = \left[\mathbf{I} - \mathbf{K}(k)\bar{\mathbf{H}}(k)\right]\mathbf{P}(k\,|\,k-1) \tag{22}$$

The proposed PAT-EKF algorithm requires the number of $7LM^2+16L^2M+LMK$ real multiplications, whereas the Park's and Kong's algorithms require, respectively, the numbers of $3LM^2+K(3L^2+LM)$ and $5M^3+10LM^2+8L^2M+LMK$ real multiplications (K is the number of snapshots). Table 1 shows the comparison of computational complexity among these algorithms for M=3, L=8 and different number of snapshots. It is evident that the PAT-EKF algorithm has lower computational complexity than the Park's algorithm for $K \geq 30$, where $K \geq 30$ is often needed for acceptable tracking performance. Although the computational complexity is higher than the Kong's algorithm, the proposed method has much better performance as demonstrated by the simulations.

Algorithm		PAT-EKF	Park's	Kong's
Number of real multiplications	K=1	3600	432	2415
	K=10	3816	2376	2631
	K=30	4296	6696	3111
	K=50	4776	11016	3591

Table 1. Computational complexity comparison for M=3, L=8 and different K values.

2.3 PAT-EKF algorithm for tracking targets in 3-D space

The PAT-EKF algorithm is now extended to track narrow-band targets in 3-D space, where the system of sensor array and source configurations is shown in Figure 2, where $s_m(k)$ is the signal transmitted by the mth target, of which φ_m and θ_m are the azimuth and elevation respectively. ρ_m is the range from the mth target to the first (reference) sensor in the uniform linear array. As explained later, the number of sensors L must satisfy the condition $L \geq 3M+1$, where M is the number of targets. All the targets can be located in the near field or far field. In the following formulations, far-field targets are treated. The output of the lth sensor for the kth sampling interval can be expressed as

$$r_l(k) = \sum_{m=1}^{M} s_m(k - \tau_{lm}(k)) + n_l(k) \qquad l = 1,2,\cdots,L \tag{23}$$

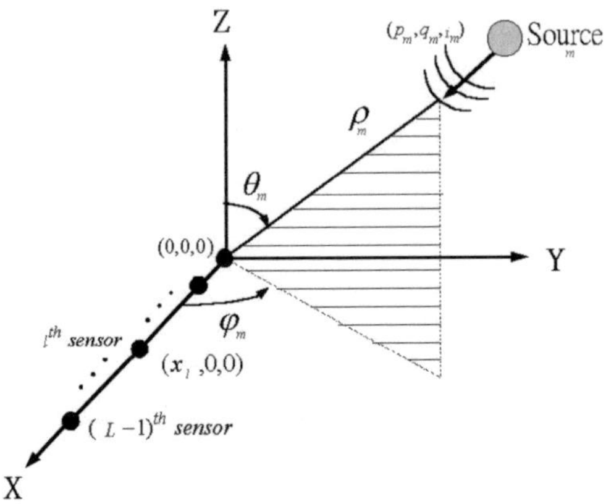

Fig. 2. Sensor array and source configurations in 3-D space.

From the array and source configurations shown in Figure 2, $\tau_{lm}(k)$ can be expressed as

$$\tau_{lm}(k) = \frac{1}{c}\left[\sqrt{(p_m - x_l)^2 + (q_m - y_l)^2 + (i_m - z_l)^2} - \rho_m\right]$$

where (x_l, y_l, z_l) is the lth sensor position relative to the reference sensor. Here $x_l=(l-1)d$, $y_l=0$, and $z_l=0$. The location coordinate of the mth signal source is given by

$$p_m = \rho_m \sin\theta_m \cos\varphi_m$$

$$q_m = \rho_m \sin\theta_m \sin\varphi_m$$

$$i_m = \rho_m \cos\theta_m$$

Assume that all the signal sources are narrowband with a common angular frequency ω. Then

$$s(k - \tau_{lm}(k)) \simeq s(k)e^{-j\omega\tau_{lm}(k)}$$

Therefore, (23) becomes

$$r_l(k) = \sum_{m=1}^{M} e^{-j\omega\tau_{lm}(k)} s_m(k) + n_l(k) \qquad l = 1, 2, \cdots, L$$

In vector-matrix notation, the received output vector of the sensor array is $\mathbf{r}(k) = \mathbf{A}(k)\mathbf{s}(k) + \mathbf{n}(k)$, where

$$\mathbf{A}(k) = \begin{bmatrix} 1 & 1 & \cdots & 1 \\ \gamma_{21}(k) & \gamma_{22}(k) & \cdots & \gamma_{2M}(k) \\ \vdots & \vdots & \ddots & \vdots \\ \gamma_{L1}(k) & \gamma_{L2}(k) & \cdots & \gamma_{LM}(k) \end{bmatrix}$$

and

$$\gamma_{lm}(k) = e^{-j\omega\tau_{lm}(k)}$$

For the use of the EKF, it redefines the state vector for the mth target as $\mathbf{x}_m(k) = [\rho_m(k) \;\; \dot{\rho}_m(k) \;\; \varphi_m(k) \;\; \dot{\varphi}_m(k) \;\; \theta_m(k) \;\; \dot{\theta}_m(k)]^T$ in 3-D space. By augmenting the real and imaginary parts of each complex matrix $\mathbf{H}_m(k)$, it has the composite real matrix of dimension $2L\times6M$

$$\bar{\mathbf{H}}(k) = \begin{bmatrix} real(\mathbf{H}_1(k),...,\mathbf{H}_M(k)) \\ imag(\mathbf{H}_1(k),...,\mathbf{H}_M(k)) \end{bmatrix}$$

which can be expressed as

$$\bar{\mathbf{H}}(k) = \begin{bmatrix} 0 & 0 & 0 & 0 & 0 & 0 & \cdots & 0 & 0 \\ g_{21}\frac{\partial\tau_{21}}{\partial\rho_1} & 0 & g_{21}\frac{\partial\tau_{21}}{\partial\varphi_1} & 0 & g_{21}\frac{\partial\tau_{21}}{\partial\theta_1} & 0 & \cdots & g_{2M}\frac{\partial\tau_{2M}}{\partial\theta_M} & 0 \\ \vdots & \vdots & \vdots & \vdots & \vdots & \vdots & \ddots & \vdots & \vdots \\ g_{L1}\frac{\partial\tau_{L1}}{\partial\rho_1} & 0 & g_{L1}\frac{\partial\tau_{L1}}{\partial\varphi_1} & 0 & g_{L1}\frac{\partial\tau_{L1}}{\partial\theta_1} & 0 & \cdots & g_{LM}\frac{\partial\tau_{LM}}{\partial\theta_M} & 0 \\ 0 & 0 & 0 & 0 & 0 & 0 & \cdots & 0 & 0 \\ c_{21}\frac{\partial\tau_{21}}{\partial\rho_1} & 0 & c_{21}\frac{\partial\tau_{21}}{\partial\varphi_1} & 0 & c_{21}\frac{\partial\tau_{21}}{\partial\theta_1} & 0 & \cdots & c_{2M}\frac{\partial\tau_{2M}}{\partial\theta_M} & 0 \\ \vdots & \vdots & \vdots & \vdots & \vdots & \vdots & \ddots & \vdots & \vdots \\ c_{L1}\frac{\partial\tau_{L1}}{\partial\rho_1} & 0 & c_{L1}\frac{\partial\tau_{L1}}{\partial\varphi_1} & 0 & c_{L1}\frac{\partial\tau_{L1}}{\partial\theta_1} & 0 & \cdots & c_{LM}\frac{\partial\tau_{LM}}{\partial\theta_M} & 0 \end{bmatrix}$$

with the following equations

$$g_{lm} = -\omega\sin(\omega\tau_{lm})\, s_m(k)$$

$$c_{lm} = \omega\cos(\omega\tau_{lm})\, s_m(k)$$

$$\begin{aligned} \frac{\partial\tau_{lm}}{\partial\rho_m} &= \frac{1}{c}\left\{[\rho_m\sin\theta_m\cos\varphi_m-(l-1)d]^2+(\rho_m\sin\theta_m\sin\varphi_m)^2+(\rho_m\cos\theta_m)^2\right\}^{-\frac{1}{2}} \\ &\quad\times[\rho_m-(l-1)d\sin\theta_m\cos\varphi_m]-\frac{1}{c} \\ &= \tau_{lm}\times[\rho_m-(l-1)d\sin\theta_m\cos\varphi_m]-\frac{1}{c} \end{aligned}$$

$$\begin{aligned} \frac{\partial\tau_{lm}}{\partial\varphi_m} &= \frac{1}{c}\left\{[\rho_m\sin\theta_m\cos\varphi_m-(l-1)d]^2+(\rho_m\sin\theta_m\sin\varphi_m)^2+(\rho_n\cos\theta_n)^2\right\}^{-\frac{1}{2}} \\ &\quad\times(l-1)d\rho_m\sin\theta_m\sin\varphi_m \\ &= \tau_{lm}\times(l-1)d\rho_m\sin\theta_m\sin\varphi_m \end{aligned}$$

$$\frac{\partial \tau_{lm}}{\partial \theta_m} = \frac{1}{c}\left\{\left[\rho_m \sin\theta_m \cos\varphi_m - (l-1)d\right]^2 + \left(\rho_m \sin\theta_m \sin\varphi_m\right)^2 + \left(\rho_m \cos\theta_m\right)^2\right\}^{-\frac{1}{2}}$$
$$\times -(l-1)d\rho_m \cos\theta_m \cos\varphi_m$$
$$= \tau_{lm} \times -(l-1)d\rho_m \cos\theta_m \cos\varphi_m$$

For the 3-D PAT-EKF algorithm, the recursive equations of (9)-(13), (16) and (20)-(22) remain unchanged and the recursive equations of (14)-(15) and (17)-(19) need to be changed as stated in the following context.

Let $\delta\rho(k)$, $\delta\varphi(k)$, and $\delta\theta(k)$ be the unknown innovations of $\rho(k)$, $\varphi(k)$, $\theta(k)$ respectively, from time k-1 to time k. The (l,m) element of $\delta\mathbf{A}(k)$ can be derived as

$$[\delta\mathbf{A}(k)]_{(l,m)} = -j\omega\gamma_{lm}(k)\left[\frac{\partial \tau_{lm}}{\partial \rho_m}\delta\rho_m + \frac{\partial \tau_{lm}}{\partial \varphi_m}\delta\varphi_m + \frac{\partial \tau_{lm}}{\partial \theta_m}\delta\theta_m\right]$$

and the residual array output with the first row removed can be expressed as

$$\Delta\tilde{\mathbf{r}}(k) = \mathbf{B}\cdot\begin{bmatrix}\delta\boldsymbol{\rho}(k)\\ \delta\boldsymbol{\varphi}(k)\\ \delta\boldsymbol{\theta}(k)\end{bmatrix} + \tilde{\mathbf{n}} \tag{24}$$

where the matrix $\mathbf{B}$ is a $(L-1) \times 3M$ matrix, given by

$$-j\omega\begin{bmatrix}\gamma_{21}\frac{\partial \tau_{21}}{\partial \rho_1}s_1 & \cdots & \gamma_{2N}\frac{\partial \tau_{2M}}{\partial \rho_M}s_M & \gamma_{21}\frac{\partial \tau_{21}}{\partial \varphi_1}s_1 & \cdots & \gamma_{2M}\frac{\partial \tau_{2M}}{\partial \varphi_M}s_M & \gamma_{21}\frac{\partial \tau_{21}}{\partial \theta_1}s_1 & \cdots & \gamma_{2M}\frac{\partial \tau_{2M}}{\partial \theta_M}s_M\\ \vdots & \ddots & \vdots & \vdots & \ddots & \vdots & \vdots & \ddots & \vdots\\ \gamma_{L1}\frac{\partial \tau_{L1}}{\partial \rho_1}s_1 & \cdots & \gamma_{LM}\frac{\partial \tau_{LM}}{\partial \rho_M}s_M & \gamma_{L1}\frac{\partial \tau_{L1}}{\partial \varphi_1}s_1 & \cdots & \gamma_{LM}\frac{\partial \tau_{LM}}{\partial \varphi_M}s_M & \gamma_{L1}\frac{\partial \tau_{L1}}{\partial \theta_1}s_1 & \cdots & \gamma_{LM}\frac{\partial \tau_{LM}}{\partial \theta_M}s_M\end{bmatrix}$$

Thus, a modified least-squares solution of (24) yields the innovations $\delta\boldsymbol{\rho}(k)$=[$\delta\rho_1(k),\cdots,\delta\rho_M(k)]^T$, $\delta\boldsymbol{\varphi}(k)$=[$\delta\varphi_1(k),\cdots,\delta\varphi_M(k)]^T$, and $\delta\boldsymbol{\theta}(k)$=[$\delta\theta_1(k),\cdots,\delta\theta_M(k)]^T$, given by

$$\begin{bmatrix}\delta\boldsymbol{\rho}(k)\\ \delta\boldsymbol{\varphi}(k)\\ \delta\boldsymbol{\theta}(k)\end{bmatrix} = (\mathbf{B}^H\mathbf{B} + \mathbf{L})^{-1}\mathbf{B}^H\Delta\tilde{\mathbf{r}}$$

These innovations are then used to update the state estimation according to

$$\hat{\theta}_m(k) = \hat{\theta}_m(k\,|\,k-1) + \delta\theta_m(k)$$

$$\hat{\varphi}_m(k) = \hat{\varphi}_m(k\,|\,k-1) + \delta\varphi_m(k)$$

$$\hat{\rho}_m(k) = \hat{\rho}_m(k\,|\,k-1) + \delta\rho_m(k)$$

There is one limiting condition, i.e., $L-1 \geq 3M$, under which the $L-1$ linear equations are used for solving $3M$ unknown variables.

3. Simulation results and discussion

In this section, the tracking performance of the three tracking algorithms are compared for narrow-band sources in 2-D space. A uniform linear array of eight sensors L=8 with half wavelength as the inter-element spacing is used. Three moving targets on the plane are tracked over an interval of 180s with T=1 s. During each T interval, K(=1, 10, 30, 50) snapshots of sensors data are generated. For comparison, the algorithms developed by Park *et al.* (1994), Kong and Chun (2000) were simulated. The Monte Carlo simulations of 100 runs were carried out for each algorithm with various SNRs. The parameters used in the system model for all algorithms to be compared are σ_v^2 =3, σ_w^2 =1, and σ_n^2 =3. The weighting factors to constrain the absolute values of innovations in (18) are set to be $l_m = \frac{1}{20} \times$ (mth diagonal element of $\mathbf{B}^H\mathbf{B}$), which is the same as in Park's algorithm (Park *et al.*, 1994). The SNR is defined as $10log(s/\sigma_n^2)$ in dB, where s is the signal power.

Table 2 gives the tracking results for various SNRs at K=30 snapshots. The PAT-EKF algorithm shows the highest tracking success rate (true angle ±5°) for each SNRs. Table 3 presents the tracking results for various number of snapshots at SNR=10dB. Again, the proposed algorithm shows the highest tracking success rate for each number of snapshots.

SNR (dB)	Tracking success rate (%)		
	PAT-EKF	Park's	Kong's
0	28	14	11
5	62	44	34
10	86	62	60

Table 2. Tracking performance for various SNRs at K=30

Number of Snapshots	Tracking success rate (%)		
	PAT-EKF	Park's	Kong's
1	70	45	43
10	83	69	55
50	88	81	66

Table 3. Tracking performance for various number of snapshots at SNR=10dB

Figure 3 shows typical sample run for crossing tracks, all based on a single snapshot of data vector (K=1) at SNR=10dB of each target. The PAT-EKF and Park's algorithms exhibit much better tracking capability than Kong's algorithm (Kong & Chun, 2000) especially at the cross points in the trajectory.

Two moving targets are tracked over an interval of 20s with T=1 in 3-D space. During each T interval, K(=160) snapshots of sensors data are generated. Figure 4 shows the tracking performances of the 3-D PAT-EKF algorithm for the combinations of range, elevation, and azimuth at 3dB of SNR. In Figure 4, dot represents the true angle and line represents the tracked angle. The 3-D PAT-EKF algorithm is very effective in tracking the targets in 3-D space, even at low SNR.

Note that the PAT-EKF algorithm is excluded for performance comparison, simply because it fails to track the angle of the first signal source for the simulation example illustrated in Figure 5. In this example, the trajectory of the first signal source reveals its significantly changing behavior of angles. This also indicates that tracking capability of the PAT-EKF algorithm is rather limited when there exists a signal source with significant rates of angle variation.

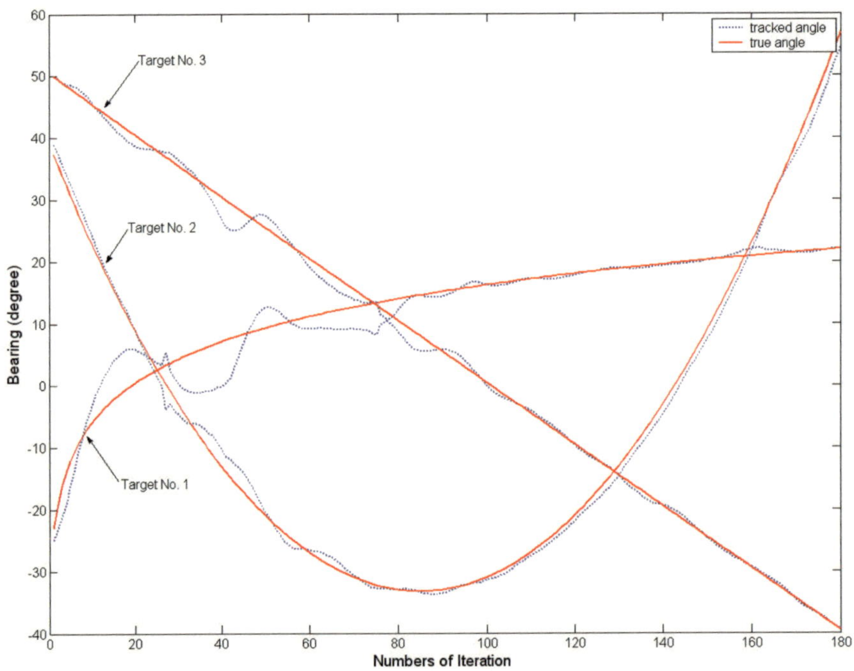

(a)

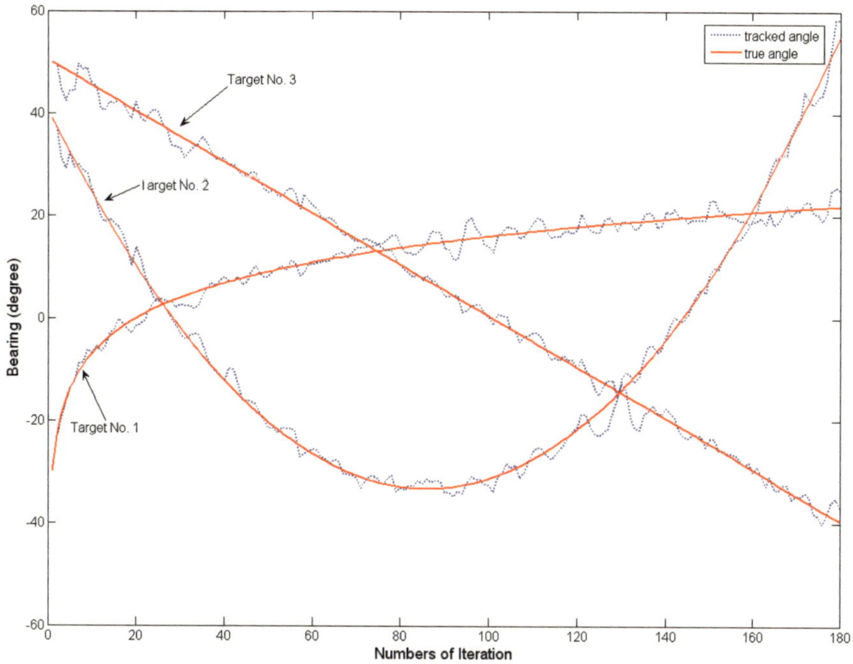

(b)

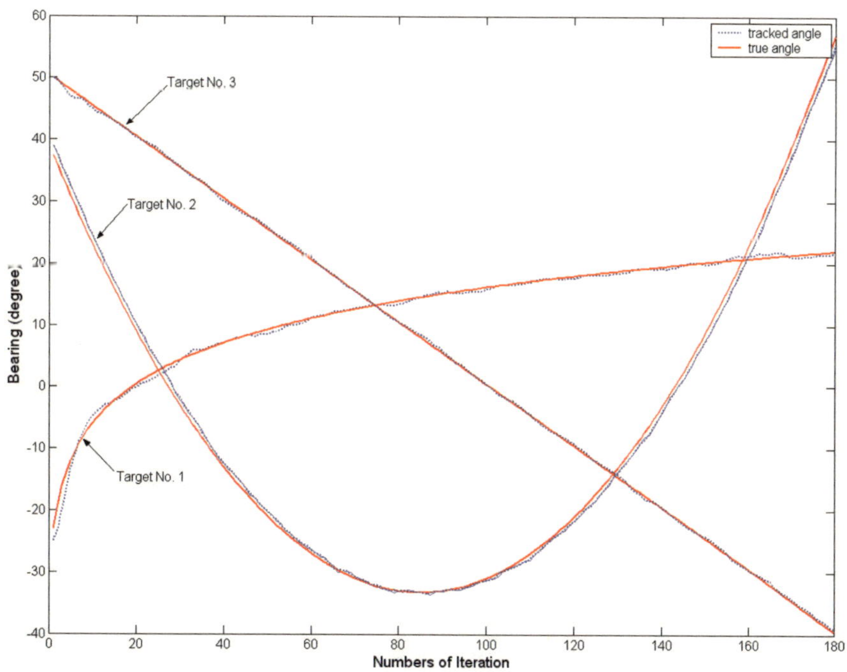

(c)

Fig. 3. Typical sample run for crossing tracks with three targets at SNR=10dB. (a) Kong's algorithm, (b) Park's algorithm, (c) PAT-EKF algorithm

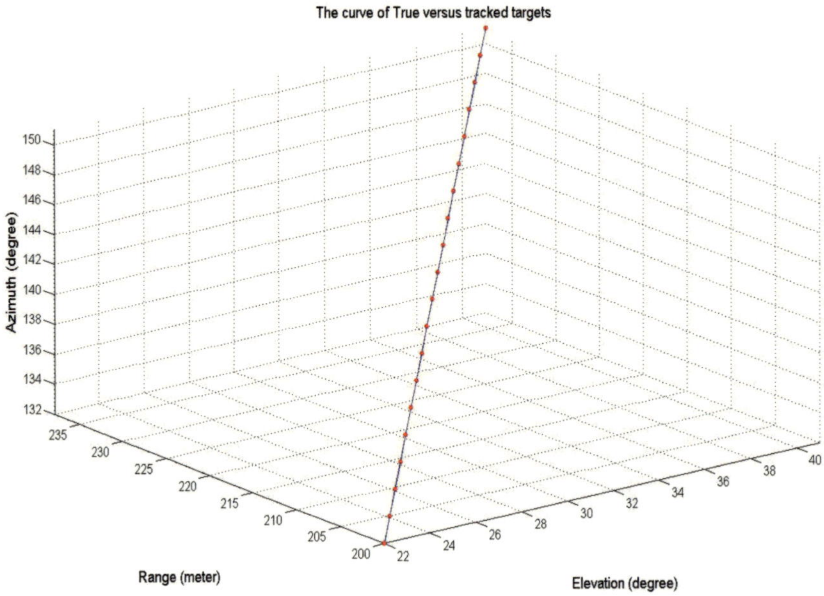

Fig. 4. Typical sample run for crossing tracks with three targets at SNR=10dB. (a) Kong's algorithm, (b) Park's algorithm, (c) PAT-EKF algorithm

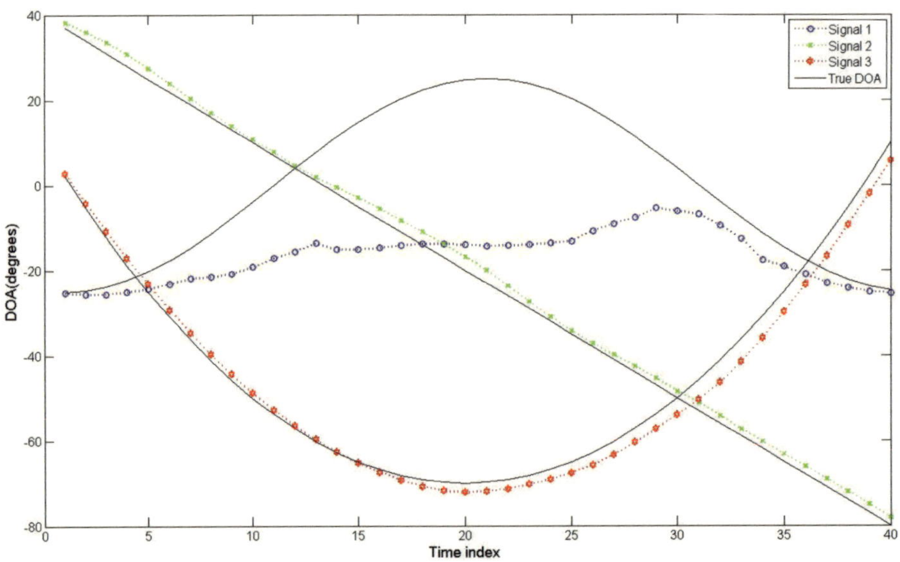

Fig. 5. The averaged tracking trajectories, using the PAT-EKF algorithm, for three equi-powered moving sources based on 25 snapshots at SNR=10dB.

4. Concluding remarks

This chapter has presented the PAT-EKF algorithm for tracking multiple targets. The proposed algorithm modified Park's algorithm by using the sensor array output vector rather than the sample covariance matrix and incorporating EKF instead of KF. These modifications allow the proposed algorithm to lower computational load, and also improve the tracking success rate particularly at lower snapshots. The PAT-EKF algorithm is then extended to track the azimuth, elevation, and range of multiple targets in 3-D space. Through computer simulations, the effectiveness of each proposed algorithm has been demonstrated. The drawback of the PAT-EKF algorithm is that it fails to track any target with a significant rate of angle variations.

5. Acknowledgment

This work was supported by National Science Council of Taiwan under contracts NSC 99-2923-E-022-001-MY3.

6. References

Chang, S.-H.; Hou, S.-Y.; Chang, S.-C. & Hung, H.-S. (2005). Underwater Wideband Signal Tracking based on Predictive Angle Tracking Algorithm. *Journal of Marine Science and Technology*, Vol. 13, No. 1, pp. 46-53

Javier, S. A. & Sylvie, M. (1999). An Efficient PASTd-Algorithm Implementation for Multiple Direction of Arrival Tracking. *IEEE Trans. Signal Processing*, Vol. 47, No. 8, pp. 2321-2324

Kong, D. & Chun, J. (2000). A fast DOA tracking algorithm based on the extended Kalman filter. Proceeding of the IEEE *National Aerospace and Electronics Conference NAECON*, pp. 235-238, ISBN 0-7803-6262-4, Dayton, Ohio, USA, Oct. 10-12, 2000

Park, S. B.; Ryu, C. S. & Lee, K. K. (1994). Multiple target tracking algorithm using predicted angles. *IEEE Trans. Aerosp. Electron Syst.*, Vol. 30, No. 2, pp. 643-648

Rao, C. R.; Sastry, C. R.; & Zhou, B. (1994). Tracking the Direction of Arrival of Multiple Moving Targets. *IEEE Trans. Signal Processing*, Vol. 42, No. 5, pp. 1133-1144

Ryu, C. S.; Lee, S. H. & Lee, K. K. (1999). Multiple target angle tracking algorithm using innovation extracted from signal subspace. *IEE Electronics Letters*, Vol. 35, No. 18, pp. 1520-1522

Ryu, C. S.; Lee, J. S. & Lee, K. K. (2002). Multiple target angle tracking algorithm with efficient equation for angular innovation. *IEE Electronics Letters*, Vol. 38, No. 10, pp. 483-484

Schmidt, R. O. (1986). Multiple emitter location and signal parameter estimation. *IEEE Trans. Antennas, Propagation*, Vol. 34, No. 3, pp. 276-280

Shan, T. J.; Wax, M. & Kailath, T. (1985). On spatial smoothing for direction -of-arrival estimator of coherent signal. *IEEE Trans. Acoust., Speech, Signal Processing*, Vol. ASSP-33, pp. 806-811

Sword, C. K.; Simaan, M. & Kamen, W. W. (1990). Multiple target angle tracking using sensor array output. *IEEE Trans. Aerosp. Electron Syst.*, Vol. 26, No. 3, pp. 367-372

Wang, H. & Kaveh, M. (1985). Coherent Signal-Subspace Processing for the Detection and the Estimation of Angles of Arrival of Multiple Wide-band Sources. *IEEE Trans. Acoust., Speech, Signal Processing*, Vol. ASSP-33, pp. 823-831

Yang, B. (1995). Projection approximation subspace tracking. *IEEE Trans. Signal Processing*, Vol. 43, pp. 95-107

Yang J.-F. & Kaveh, M. (1988). Adaptive eigensubspace algorithms for direction or frequency estimation and tracking, *IEEE Trans. Acoust., Speech, Signal Processing*, Vol. 36, pp. 241-251

4

Shape Recognition and Position Measurement of an Object Using an Ultrasonic Sensor Array

Kozo Ohtani[1] and Mitsuru Baba[2]
[1]*Hiroshima Institute of Technology*
[2]*Ibaraki University*
Japan

1. Introduction

Shape recognition of a transparent object is usually difficult to perform by image processing techniques, because the major portion of the light projected onto an object passes through the object. As a result, the object cannot gain the light intensity required for image processing. Ultrasonic sensors are often utilised in situations where such optical sensors cannot be used. Moreover, systems using ultrasonic sensors are simpler and cheaper than systems using other types of sensors. Although the ultrasonic method has such advantages, results from the conventional ultrasonic method do not always have high measuring resolution, due to the wide directional pattern of the sensing. Up to now, the application of the ultrasonic method has been limited compared to the optical method.

In an ultrasonic recognition system, ultrasonic sensors are combined with neural networks (Yoneyama *et al.*, 1988), (Farhat, 1989), (Watanabe & Yoneyama, 1990), (Masumoto *et al.*, 1993), (Serrano *et al.*, 1997). In these systems, the capability of the neural networks compensates for the low resolution of the ultrasonic method. However, in the conventional ultrasonic methods using neural networks (Holland, 1992), data used for the neural networks are given by acoustic holography, tomography or time-of-flight measurement. As a result, it is difficult for the conventional ultrasonic methods to improve the low resolution, because shape recognition using these methods has been limited to objects with simple shape. In addition, it has been difficult to measure the position of an object.

The time-of-flight method involves measuring the time arriving at an object by ultrasonic wave pulses or amplitude modulations, in which the distance information obtained is input into a neural network for measurement and recognition, as shown in Fig. 1 (b). Although this method yields high resolution in the depth direction, the width direction resolution is limited by the arrangement interval of the ultrasonic sensor array. As a result, there are measurement and recognition constraints in principle, such as the facts that it is difficult to measure the position in the width direction with high resolution and objects differing slightly in size cannot be differentiated. In contrast, the acoustic holographic method reproduces an image of an object by interfering with the scattering or penetrating waves produced by irradiating ultrasound on a measured object with reference waves, as shown in Fig. 1 (c). This method yields good resolution in the width direction and is superior in two-dimensional measurements and recognition. However, since the distance information in the depth direction in the holographic image that is obtained, it has the drawbacks that a holographic image is needed for each distance if the depth-direction distance to an object varies, and it is difficult to directly detect changes in the pose of a target object. Thus, there are a number of problems that must be resolved to achieve further progress in measurement and recognition systems based on these methods and neural networks.

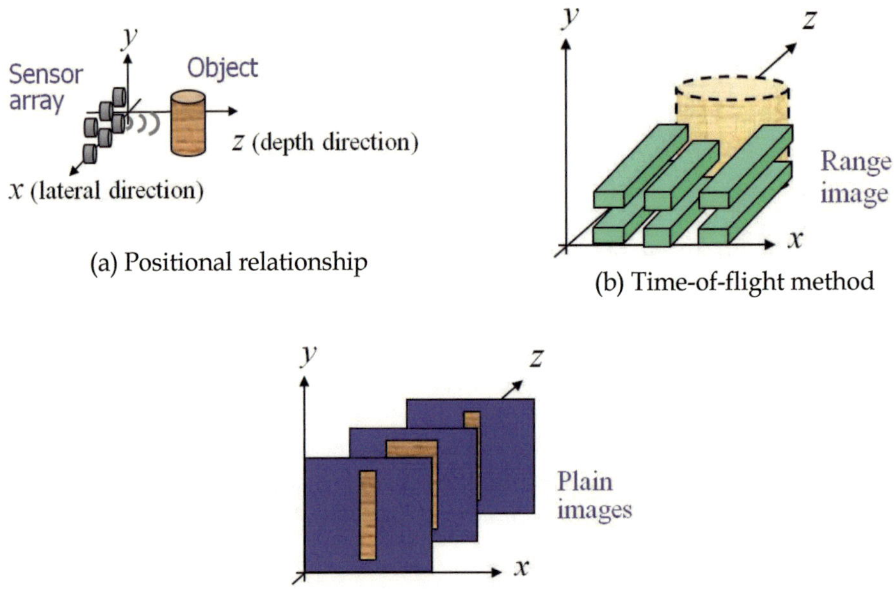

Fig. 1. Acoustic image obtained by conventional ultrasonic methods

The authors have studied an ultrasonic object position and shape recognition system that can improve the distance resolution in both depth and width directions simultaneously by directly using sound pressure signals of an ultrasonic sensor array based on viewpoints different from those of the past methods (Ohtani, 2002), (Ohtani & Baba, 2007). From the directional pattern of the ultrasonic sensor and the attenuation characteristics of ultrasonic signals, the sound pressure signals of the sensor are generated that contain information required for identifying the position in the form of sound pressure intensity and information required for object shape recognition based on the sound pressure distribution. In this chapter, the construction of a new ultrasonic recognition system for transparent objects with complex shapes is introduced, for which a commercial polyethylene terephthalate bottle (PET bottle) is employed as a model. The sensor system consists of an ultrasonic transmitter, an ultrasonic receiver consisting of an ultrasonic sensor array, and a recognition unit with neural networks. The system locates the object between the ultrasonic transmitter and the ultrasonic receiver. It can identifies the object and measures the position of the object simultaneously. In the following section, the basic principle, system configuration, and experimental results of the ultrasonic identification system is described.

2. Basic principle

2.1 Outline

Fig. 2 shows the construction of the proposed system. The system primarily consists of an ultrasonic sensor array unit, a signal processing unit, and an identification unit. In this study, the x coordinate position on the x - z plane is defined as the width direction position of the measured object, and the z coordinate position on the $x - z$ plane is defined as the depth direction position of the measured object in the coordinate system as shown in Fig. 2 The slope of the base of the object with respect to the $x - y$ plane is defined as the pose of the object. The ultrasonic sensory array unit has two-dimensional ultrasonic receivers and one transmitter. The transmitter positions the centre of the receiver's array and irradiates an ultrasonic signal to the measured object. The signal processing unit extracts the features of the measured object from the sensor array outputs using the signal processing circuit. The extracted data is inputted into the identification unit. The identification unit consists of two types of neural networks that perform shape identification and material identification, respectively.

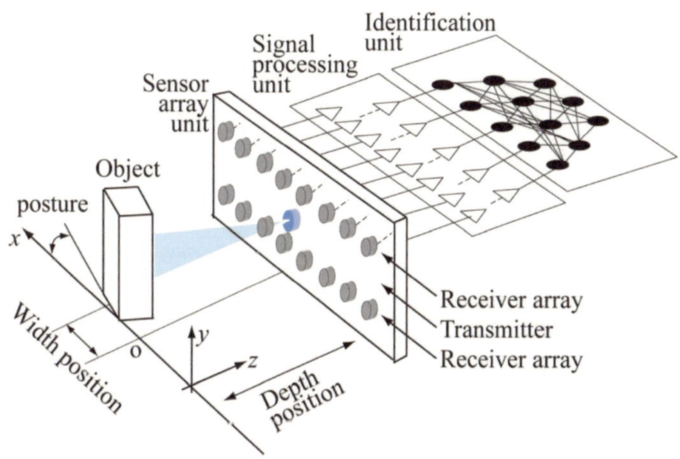

Fig. 2. System configuration of the ultrasonic sensor system

2.2 Ultrasonic wave propagation characteristics

The reflected ultrasonic pressure from a measured object is required in relation to three elements: the reflection ratio of the measured object, the distance between the object and the sensor array unit, and the directivity of the ultrasonic signal (Kocis and Figura, 1996). First, the reflection ratio relates to the acoustic impedance. The acoustic impedance, z, is defined as the product of the density and the acoustic velocity, c, of the object, as expressed by equation (1). When an ultrasonic signal propagated in Medium 1 reaches Medium 2, as shown in Fig. 3, the reflected ratio, r, and the permeation ratio, τ, at the changing point of the mediums is expressed as equations (2) and (3). In this way, the reflection ratio depends on the acoustic impedance of the object. The acoustic impedance of a measured object is correlated with the material. Therefore, the magnitude of the reflected ultrasonic pressure gives the information required for material identification.

$$z = \rho c \tag{1}$$

$$r = \frac{z_1 - z_2}{z_1 + z_2} \tag{2}$$

$$\tau = \frac{2z_2}{(z_1 + z_2)^2} \tag{3}$$

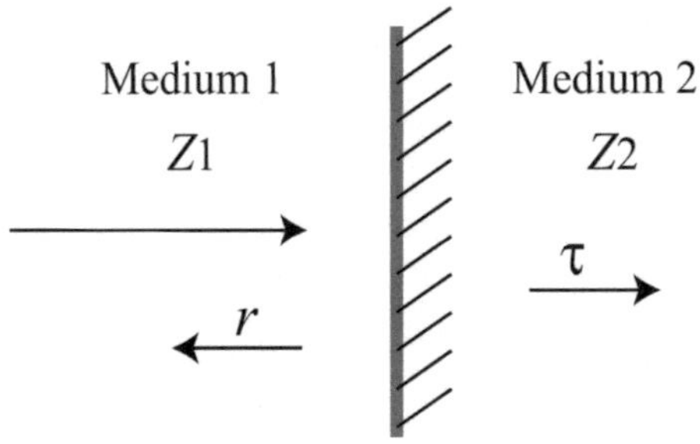

Fig. 3. Reflection and penetration

Second, the ultrasonic pressure, p, is reduced by the propagation distance, d, as expressed by equation (4). Here, a stands for the radius of a vibration board that produces an ultrasonic wave, and v_0 stands for the velocity of the vibration, as shown in Fig. 4. In this way, the reflected ultrasonic pressure from the measured object has a relation to the flight distance.

$$|p| = 2\rho c |v_0| \, |\sin\frac{\pi}{\lambda}(\sqrt{d^2 - a^2} - d)| \tag{4}$$

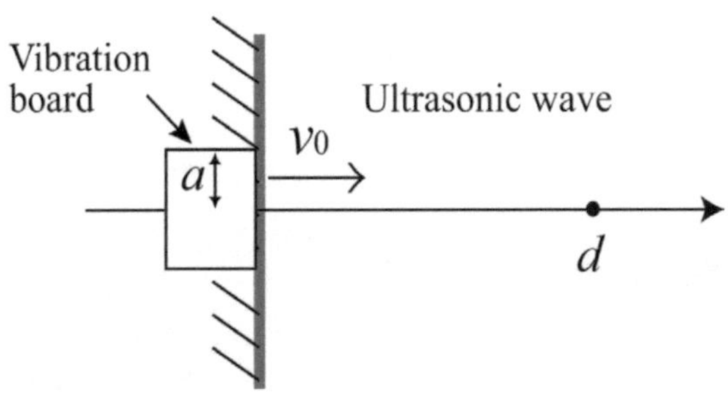

Fig. 4. Reduction by distance

Third, because of the directivity of an ultrasonic wave, the ultrasonic pressure detected by a sensor array has a relation to the shape and the orientation of the measured object. As a result, even if the positions of two objects are the same, but the pose of the objects are different from each other, the reflected ultrasonic pressure from these objects are different. Therefore, in this stage, the proposed method has a little limitation to the pose of measured objects to simplify the identification process. In these circumstances, the reflected ultrasonic pressure is regarded as the key to identifying the shape and position of measured objects.

2.3 Feature values for shape recognition and position measurement

Fig. 5 shows how the reflected waves from an object vary according to the shape of the side of the object. Fig. 5 (a) presents the output distribution of the reflected waves when the side of a rectangular prism and the ultrasonic transmitter are placed in parallel. Fig. 5 (b) and (c) present the case of a cob of corn and a triangular prism as the measured object, respectively. The edge of the triangular prism faces the transmitter. Fig. 5 (d) presents the positional relationship between the object and the sensor array. In these figures, because the size of the receiver array is 8 x 5, the output distributions are plotted by five kinds of marks. As seen in the figures, in the case of a rectangular prism, the peak value of the output distribution is larger than that of others. There are two peaks in the output distribution of a triangular prism because of the edge of the object. The area of the output distribution of a cob of corn is comparatively wide. Therefore, the peak values and the area of the amplitude output distribution of the reflected signals received by the ultrasonic sensor array are different, corresponding to the edge forms of the sides of the objects.

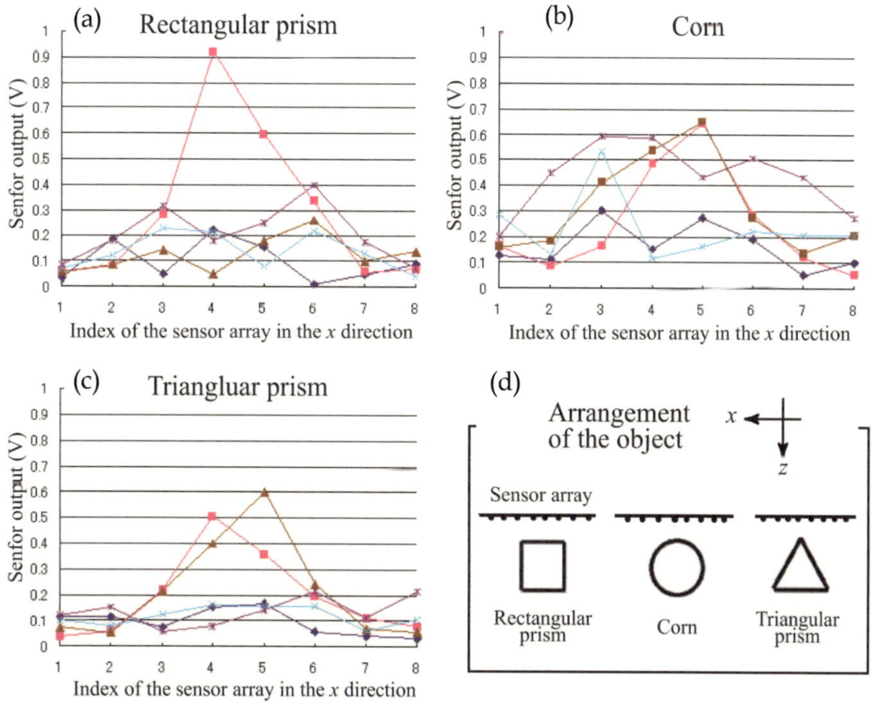

Fig. 5. Shape recognition principle

Fig. 6 shows the output distributions that differ in accordance with the position of the object. Fig. 6 (d) presents the arrangement of the object with respect to the sensor array. As shown in Fig. 6 (b), when the measured object is leaning to the right, one peak occurs in the sensor output distribution, and the peak position moves to the right. In the same way, when the measured object is leaning to the left, the peak position of the output distribution moves to the left. Moreover, as shown in Fig. 6 (c), when the measured object is leaning to the depth-direction, the peak value of the output distributions gets smaller because of the attenuation of the ultrasonic waves. Thus, the positional information of the object is contained in the peak position and the peak value.

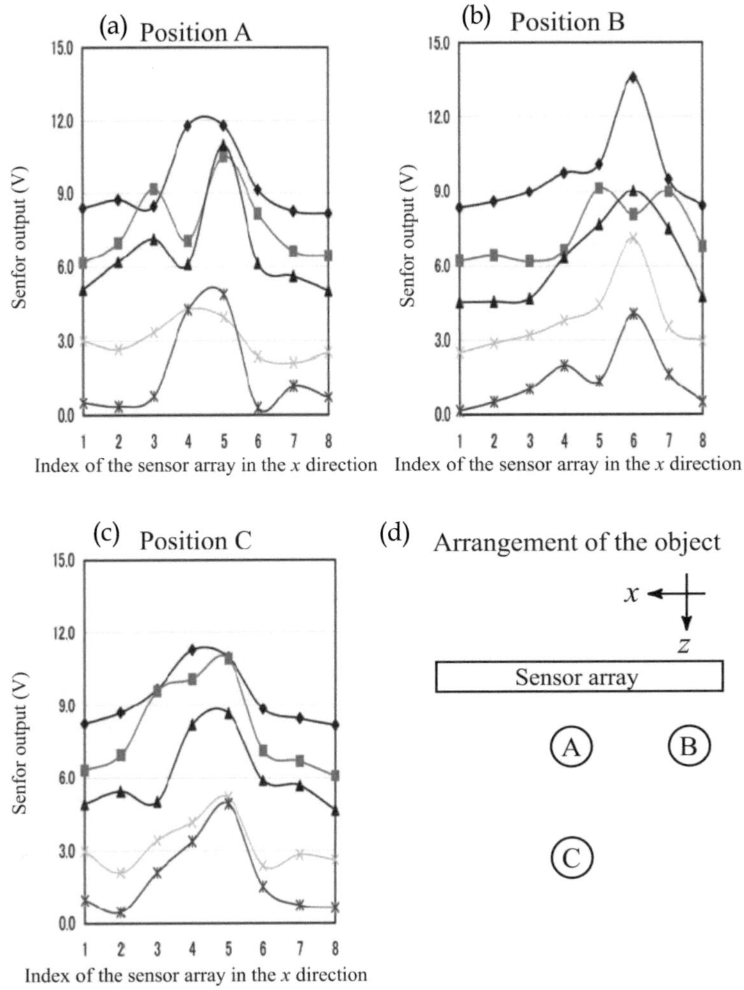

Fig. 6. Position measurement principle

In the case of PET bottles, Fig. 7 shows the received ultrasonic wave distribution of the receiver for PET bottle 1, PET bottle 2, and PET bottle 3 among 7 PET bottles, as shown in Fig. 12. These PET bottles have almost the cross section of a square on the bottom side and the cross section of a circle in the top side. In these figures, because the size of the receiver array is 8 x 2, the output distributions are plotted by two kinds of lines. Every peak value of the ultrasonic distribution of the bottom side in these three PET bottles is lower than that of the top side, because the indentation of the surface of the bottom side causes a scattering of ultrasonic waves. In this way, the proposed method identifies the shape of a PET bottle using information of the distorted ultrasonic waves. As the figure shows, the shape of the PET bottle under study can be identified by analyzing the ultrasonic wave distributions. PET bottle 5 is a hexagon, and PET bottle 6 is a cylinder.

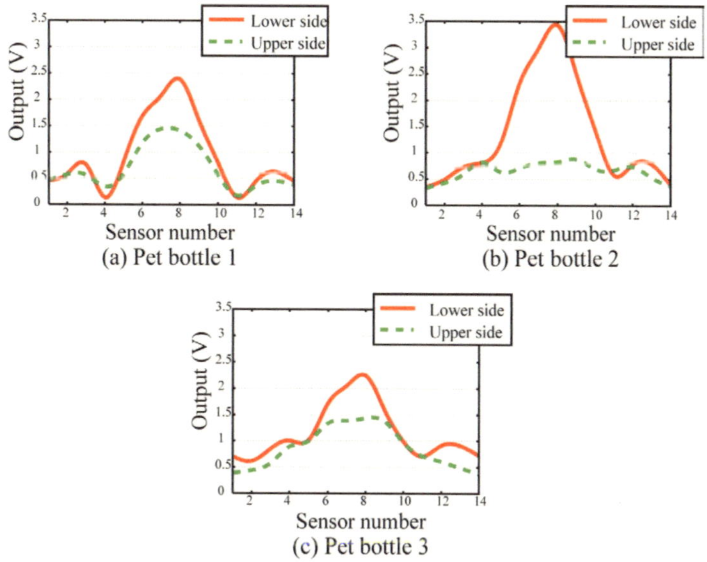

Fig. 7. Ultrasonic pressure distribution of different shapes of PET bottle

Figure 8 shows the mechanism for measuring the object's position. The positions in the x-direction and in the z-direction of a PET bottle have been measured by moving in parallel with and vertical with respect to the sensor array, respectively, as shown in Fig. 8 (a). Fig. 8 (b) shows the ultrasonic pressure distribution when the PET bottle is shifted in the x-direction. Fig. 8 (c) shows the distribution when the PET bottle is moved in the z-direction. From these figures, it is shown that the positions and values of the peaks of the ultrasonic pressure distribution change as the PET bottle moves. In this way, the proposed method uses the values and locations of the wave peaks of the ultrasonic pressure distribution to perform position measurement of the PET bottles.

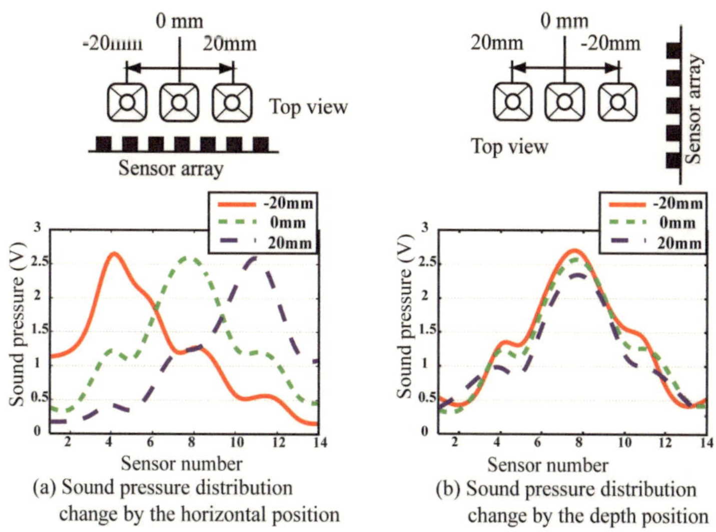

Fig. 8. Mechanism of position measurement for PET bottles

2.4 Recognition and position measurement by a neural network

Neural networks perform recognition and position measurement of a target object. Fig. 9 shows the neural network used for the experiment, which is a five-layered feed-forward neural network consisting of one input layer, three hidden layers, and one output layer. The input layer is composed of 28 input units. The data inputted to the input units include all outputs of the sensor array of the upper side and bottom side, each of which has 14 ultrasonic sensors. The hidden layer has 3 x 15 hidden units. The output units are given corresponding to the number of objects of shape recognition. In the following experiment, the output layer has 8 units, which includes 7 output units for recognition of the target object and 1 output unit for position measurement.

The teaching pattern data for output units is given as follows. In the teaching mode of the neural network, let the unit for the target object be 1 and the others be 0 for shape recognition units. For the position measurement unit, let the unit be from 0 to 1 according to the normalised range to be measured. Therefore, in the recognition mode and the measurement mode, a recognition result is evaluated using the ratio of the output for the measured object to the amount of all the outputs. Successful recognition is regarded as a ratio of over 50%.

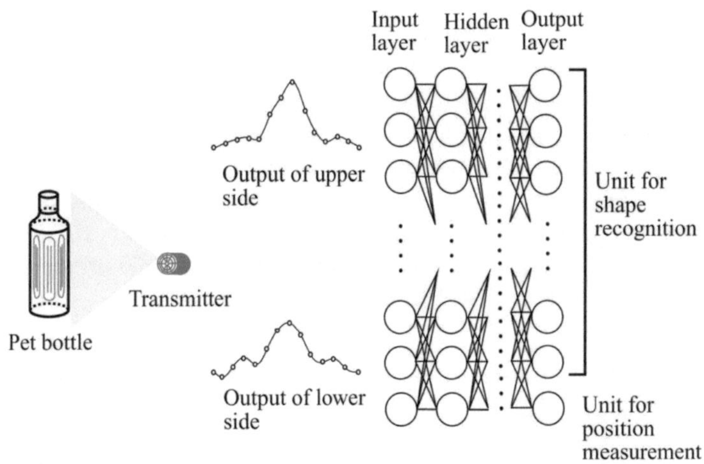

Fig. 9. Construction of the neural network

3. Experiments

3.1 Setup

Fig. 10 shows the experimental setup, which consists of a transmitter, a receiver, and a measured object. The transmitter, which operates at 40 kHz and approximately 120 dB output, is located between the sensor array of the upper side and the sensor array of the lower side. Ultrasonic waves have a wider directivity as the frequency decreases, and signals reflected by objects other than the measured object of the experimental equipment and the like are frequently input as noise to the ultrasonic sensor array, in addition to the signals directed onto the target object. Since an ultrasonic sensor is used to transmit a 40 kHz signal with a comparatively broad directive angle, considering the sizes of the sensor elements and the measurement range of the ultrasonic sensor array, the effects of such noise are pronounced. To alleviate such effects, a cylindrical cover is attached as shown in Fig. 10.

The upper and lower side each have a receiver and two sensor arrays. Each sensor array is composed of 14 ultrasonic sensors that are arranged linearly at about 15 mm intervals. The ultrasonic sensor used for the sensor array is a received sensitivity of -64 dB. The ultrasonic wave received by the receiver is amplified by an operational amplifier (op-amp), and the sinusoidal wave of the ultrasonic transmission is held by a peak hold circuit. The gain of the op-amp is controlled to reduce the variation of the specific characteristics of receivers. Fig. 11 shows the effectiveness of the gain control. Fig. 11 (a) shows the output distributions without the gain control, and Fig. 11 (b) shows the output distributions with the gain control. The output distribution with the gain control is almost symmetrically unimodal, and the variations are reduced. An analogue-to-digital converter (ADC) is connected to the output terminal of each receiver.

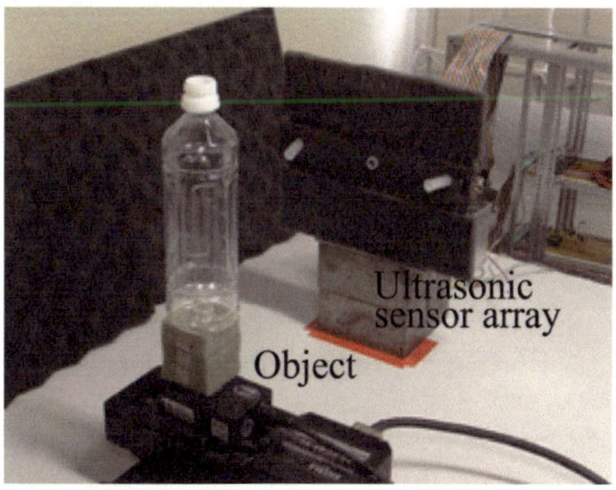

Fig. 10. Photo of the experimental setup

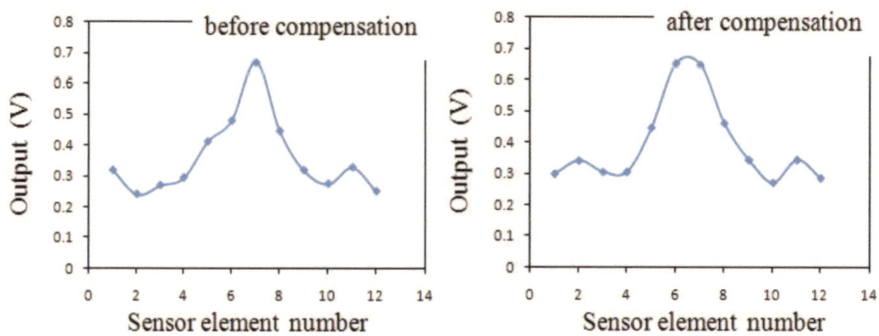

(a) Before compensation (b) After compensation

Fig. 11. Compensation of the sensor specific characteristics

3.2 Measured objects

Fig. 12 shows the objects used for the experiment. These are 7 kinds of transparent plastic commercial PET bottles, which are filled with coloured liquid so that their shape can be easily discerned. The PET bottles used in the experiment have the right rectangular prism form (PET bottle 1 - PET bottle 4), hexagonal column form (PET bottle 5 and PET bottle 6), and cylinder form (PET bottle 7). PET bottle 4 is shorter than the other bottles. Although the cross sections of these PET bottles have almost the same shape, the indentation of the surface is somewhat different. PET bottle 5 and PET bottle 6 have the same cross section at the bottom side of the bottle. However, they are different in the cross section at the top side of the bottle, that is, PET bottle 5 is a hexagon and PET bottle 6 is a cylinder.

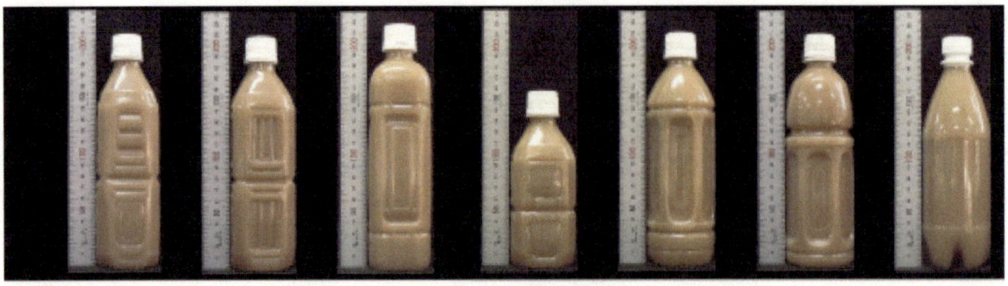

PET bottle 1 2 3 4 5 6 7

Fig. 12. Measured objects

3.3 Experimental procedure

The experimental procedure was as follows:

1. One PET bottle was placed in front of the transmitter and the receiver. The distance between the transmitter (the receiver) and the PET bottle was about 130 mm, as shown in Fig. 13.
2. The object moved in the x-direction in 2 mm increments to acquire the data to be used in the neural networks in the teaching mode. In this way, teaching data were acquired for 21 patterns.
3. In this experiment, there are 7 output units of the neural networks for the object identification, because the recognition system needs to identify 7 kinds of PET bottle. If the measured PET bottle is PET bottle 1, the system is taught in such a manner that the output unit 1 is nearly 1, while the outputs of the other units are nearly 0. Similarly, if the measured PET bottle is PET bottle 2, the output unit 2 is nearly 1, and those of the other units are nearly 0.

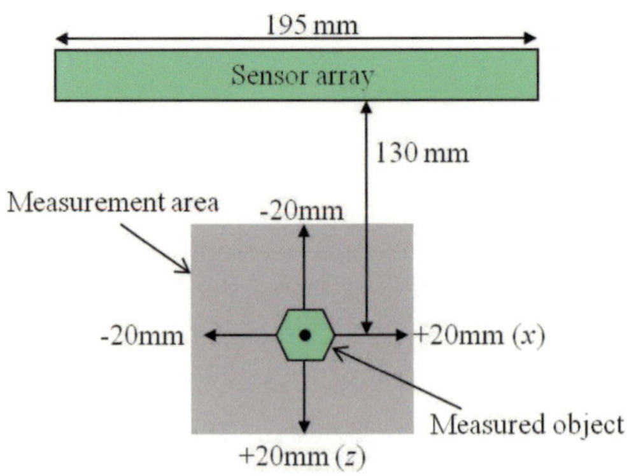

Fig. 13. Experimental conditions

4. The number of the output unit of the neural network for the position measurement is one, and the neural network outputs the analogue values of the positions of the measured PET bottle in the x-direction or the z-direction.
5. These steps were repeated to teach the neural network until the squared error became smaller than 0.00001 for every pattern in the teaching mode.
6. (6) Data were acquired for the 20 patterns used in the experimental mode for the PET bottle in question.
7. (7) These steps were repeated for all 7 PET bottles.

3.4 Results

Table 1 shows the recognition results in the case of free position of an object. The recognition was perfect for all PET bottles, indicating that the method has the ability to recognize a PET bottle.

Object	Trial	Sucess
PET bottle 1	20	20
PET bottle 2	20	20
PET bottle 3	20	20
PET bottle 4	20	20
PET bottle 5	20	20
PET bottle 6	20	20
PET bottle 7	20	20

Table 1. Shape recognition results

Fig. 14 shows the linearity of three typical PET bottles in the x-direction and the z-direction, respectively. The actual positions are plotted along the vertical axis, and the values outputted by the neural network are plotted along the vertical axis. The maximum position error in the x-direction was about 3.21 mm and in the z-direction was about 5.46 mm. Table 2 lists the maximum position error of all measured PET bottles in the x-direction and the z-direction. From Table 2, it can be seen that the measurement accuracy of the position in the z-direction became worse compared to the measurement accuracy of the position in the x-direction. The reason is that although the positions in the z-direction were measured based on peak values of the ultrasonic pressure distribution, the change of the peak value was small in the present experiment.

Object	Maximum error(mm)	
	x	y
PET bottle 1	2.02	2.46
PET bottle 2	1.25	2.65
PET bottle 3	1.58	3.18
PET bottle 4	0.96	2.89
PET bottle 5	1.15	2.95
PET bottle 6	2.03	5.46
PET bottle 7	3.21	2.86

Table 2. Position measurement results

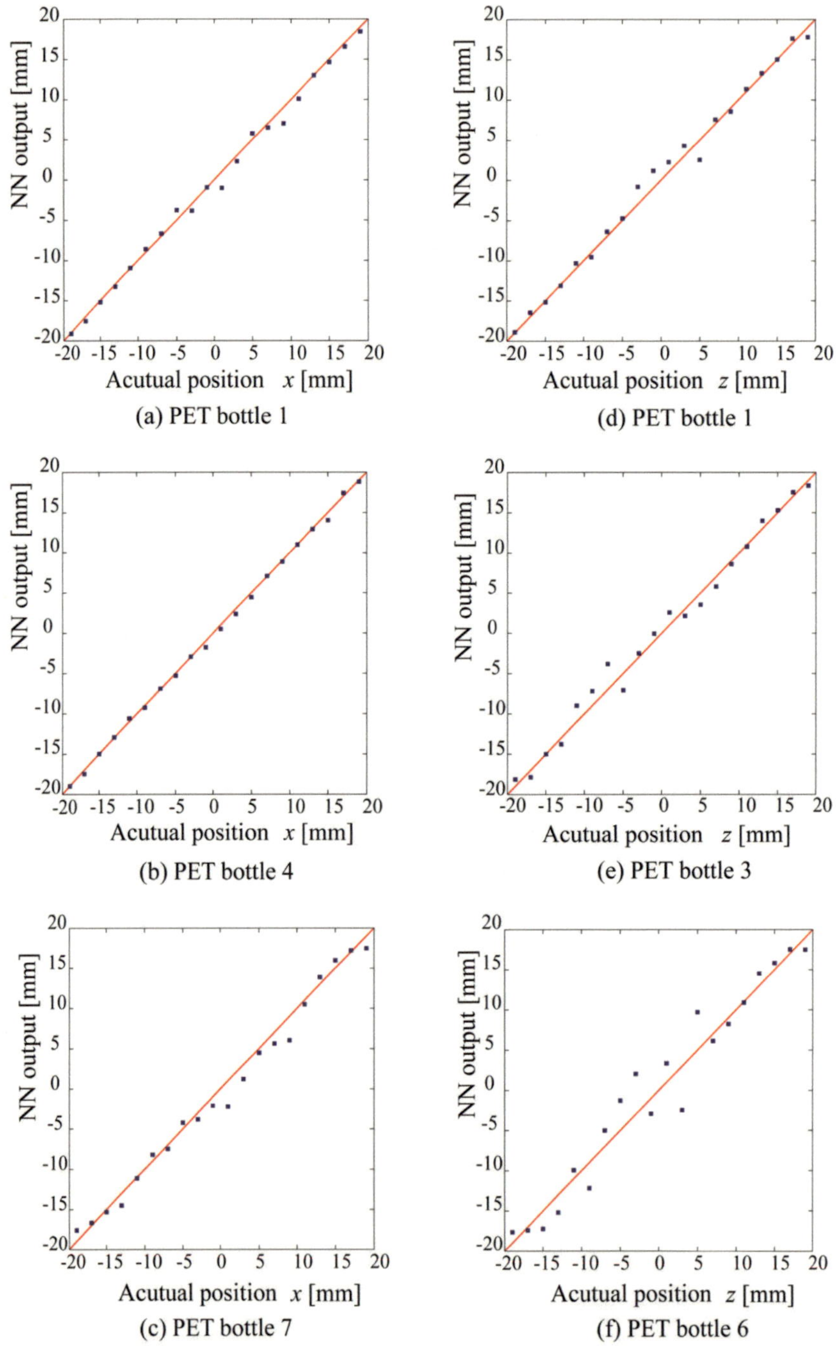

Fig. 14. Position measurement results in x-direction (a)~(c) and in z-direction (e)~(g)

The experimental results indicate that the sensor system is effective for industrial applications that identify and measure the position of a transparent object such as a PET bottle or a glass bottle.

3.5 Other experimental results

Another experiment was performed with the prototype sensor system, in which a sphere, rectangular prism, cylinder, triangular prism, regular pyramid, triangular pyramid, and spherical pyramid, as shown in Fig. 15 were measured. The measurement procedure was as follows. First, the teaching pattern data for the neural networks were obtained in 3 mm intervals in the z direction. The distance between the sensor array and the object was 200-290 mm. Next, learning was performed by the neural network over the appropriate distance. In this experiment, the neural network required the teaching patterns obtained in the range of 6 mm in the z direction in consideration of the error of the distance measurement. Finally, an object was measured as a test pattern. 31 measurement patterns were used for shape identification.

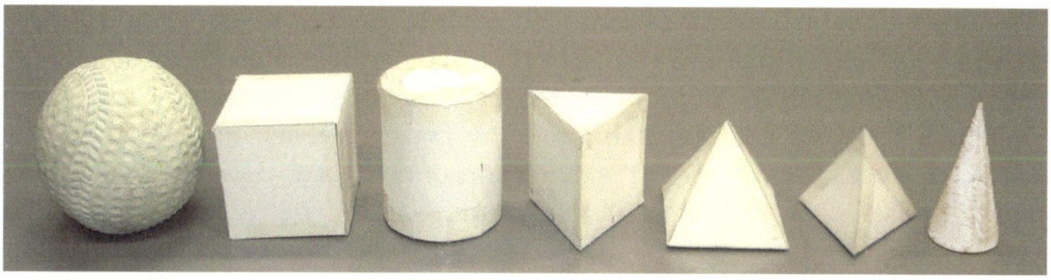

Fig. 15. Measured objects

Table 3 shows the results of shape identification. In this experiment, the pose of the object was fixed, and the number of test patterns was 31. As the table indicates, all the identifications were successful.

Object	Trial	Success
Cylinder	31	31
Triangular prism	31	31
Rectangular prism	31	31
Spherical pyramid	31	31
Triangular pyramid	31	31
Regular pyramid	31	31
Sphere	31	31

Table 3. Shape recognition results

3.6 Discussion

The results of the experiments verify that high-resolution position measurements and shape recognition are possible by the method proposed in this study with respect to PET bottles. That is, the proposed scheme is a method that can obtain high-resolution distance information in both the depth and width directions, which is the original goal of the scheme, and that this scheme is a promising method for ultrasonic position and shape recognition. The method in this study is limited to the illustrated object orientation in its current stage. To relax the limitations on the measurements and recognition, it is necessary to discuss the arrangement of the sensor array and the irradiation direction of the ultrasonic waves with respect to the measured objects.

4. Conclusions

In this chapter, the construction of a new ultrasonic recognition system for a transparent object using both ultrasonic sensors and a neural network is described. The proposed system consists of an ultrasonic transmitter, an ultrasonic receiver, and a recognition unit. It simultaneously can identify a PET bottle and measure the position of the PET bottle with a neural network. A prototype sensor system has been used to recognise and measure 7 kinds of PET bottle. Experimental results demonstrate that the sensor system achieved perfect recognition, and position measurement with accuracy to within 3.21 mm in the x-direction and within 5.46 mm in the z-direction. In conclusion, the proposed system is effective for applications that identify and measure the position of transparent objects such as a PET bottle or a glass bottle.

Although problems remain with the proposed scheme, such as the fact that it is limited to the current orientation of the measured objects, it is possible to use it for identifying and differentiating products on plant production lines, which indicates the practical significance of the method. To relax the constraints on the measurements and recognition and to recognise more complicated objects, the extraction of new characteristic quantities and the arrangement of the sensor array are planned for future study.

5. References

Farhat N H (1989). Microwave Diversity Imaging and Automated Target Identification Based on Models of Neural Networks, *Proceedings of the IEEE*, Vol. 77, No. 5, pp. 670-681, May, 1989

Holland J H (1992). *Adaptation in natural and artificial systems*, The MIT Press, ISBN: 0262082136, April 1992, Cambridge, MA, USA

Kocis S & Figura Z (1996). *Ultrasonic Measurements and Technologies*, Chapman & Hall, ISBN: 0412638509, 1996, Suffolk, UK

Masumoto D, Kimoto T & Nagata S (1993). A Sensory Information Processing System Using Neural Networks - Three-Dimensional Object Recognition with Sensorimotor Fusion, *Proceedings of 1993 International Conference on Neural Networks*, Vol .II, pp. 655-660, October 25-29, 1993

Ohtani K, Baba M & Konishi T (2002). Position, Pose Measurement and Recognitions for Pillar-Like Objects Using Ultrasonic Sensor Array and Neural Networks, *Systems and Computers in Japan*, Vol. 33, No. 11, pp. 27-38, October 2002

Ohtani K & Baba M (2007). Shape Recognition for Transparent Objects Using Ultrasonic Sensor Array, *Proceedings of SICE Annual Conference 2007*, pp. 1813-1818, September 17-20, 2007.

Serrano I, Lazaro A & Oria J P (1997). A New Method for Object Identification in Ultrasonic Systems Using Neural Nets, *Proceedings 1997 IEEE International Symposium on Computational Intelligence in Robotics and Automation CIRA'97*, pp. 152-157, July 10-11, 1997

Watanabe S & Yoneyama M (1990). Ultrasonic Robot Eye Using neural networks, *IEEE Transaction Ultrasonic, Ferroelectrics, Frequency Control*, Vol. 37, No. 2, pp. 141-147, 1990

Yoneyama M, Watanabe S, Kitagawa H, Okamoto T & Morita T (1988). Neural Network Recognizing 3-Dimensional Object Through Ultrasonic Scattering Waves, *Proceedings of IEEE Ultrasonic Symposium*, pp. 595-598, October 2-5, 1988

5

Detection of Alkylating Agents Using Optical, Electrical and Mechanical Means

Yoav Eichen et al.*
[1]Schulich Department of Chemistry, Technion-Israel Institute of Technology, Technion City, Haifa, Israel

1. Introduction

Alkylating agents, such as methyl iodide, dichloromethane and epichlorohydrin, are commonly used in organic synthesis as reagents and solvents. The alkylating agents, such as methyl bromide, are still used as soil steriliers, while nitrogen mustards are being used as anticancer drugs. Alkylating agents are also dangerous byproducts of the water purification process. Unfortunately, due to their ease of production and storage, those materials are also used as chemical warfare agents. Owing to their ability to react with nucleophiles in our body, most of these materials are toxic, carcinogenic and mutagenic. The combination of simple synthesis with aggressive reaction with biological tissues makes many alkylating agents perfect potential chemical warfare agents for the underdogs and thus a perfect terror inflicting weapon. Therefore, there is a clear need for simple, sensitive and informative tools for the detection and identification of such agents, especially in the gas phase.

Tailoring selective receptors for small and reactive molecules is a non-trivial and challenging goal. In addition, it is much easier to synthesise a new toxic alkylating agent than to tailor a selective receptor for its detection, rendering this approach inefficient. Previous attempts to address this challenge resulted in two basic approaches. The first approach gave up selectivity, detecting all alkylating agents in a non-specific way, relying on a colour change of a nucleophile upon reacting with an alkylating agent (Hertzog-Ronen *et al.*, 2009). The second approach focused on a limited family of alkylators by detecting a specific mass or the presence of a specific atom it contains (for example the sulphur atom in sulphur-based mustard gas). Both approaches provide only a limited detection ability with the first having the potential of too many false positive alerts and the later having a very limited scope.

One possible solution to the detection of alkylating agents is the further development of the fist approach that uses a nucleophile as a trap for alkylating agents. An efficient nucleophile,

*Yulia Gerchikov[1], Elena Borzin[1], Yair Gannot[1], Ariel Shemesh[1], Shai Meltzman[1], Carmit Hertzog-Ronen[1], Shay Tal[2], Sara Stolyarova[3], Yael Nemirovsky[3] and Nir Tessler[3]
[1]*Schulich Department of Chemistry, Technion-Israel Institute of Technology, Technion City, Haifa, Israel*
[2]*Systems Biology Department, Harvard Medical School, Boston, MA, USA*
[3]*Department of Electrical Engineering, Technion-Israel Institute of Technology, Technion City, Haifa, Israel*

such as an amine group, is expected to react with all the reactive alkylating agents as they are essentially electrophiles as shown in Fig. 1. This reaction guarantees very high sensitivity, as the binding is not reversible.

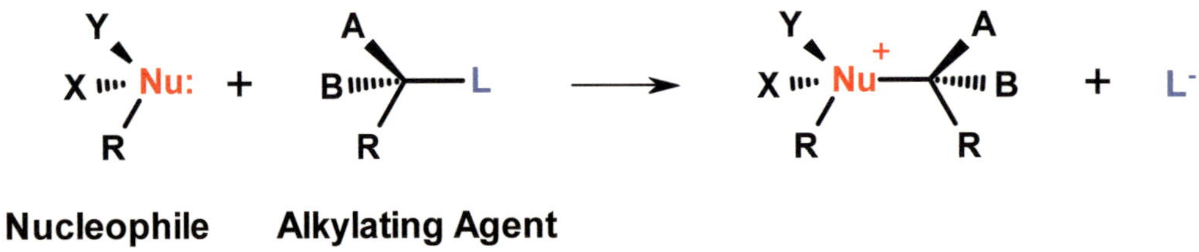

Fig. 1. General equation for reaction of a nucleophile (Nu:) with an alkylating agent

The reaction of the nucleophile with the alkylating agent needs to be transformed into a measurable signal. For example, we have recently reported on using luminescence turn-on as well as luminescence shifts for the detection and identification of different alkylating agents (Borzin *et al.*, 2010; Tal *et al.*, 2006). The signal needs to be analysed in a way that offers identification ability as well as some ability to assess the risk from the detected alkylator.

This chapter is about the coupling of the reaction between a nucleophile and alkylating agents to optical (luminescence change), electrical (organic field effect transistors, OFETs), and mechanical (micro-cantilevers) signal transducers for the detection, risk assessment and identification of alkylating agents. The three complimentary methods demonstrate the potential embedded in using such reactions for the detection of reactive substances.

2. Results and discussion

2.1 Luminescence based detection of alkylating agents

The photo-induced electron transfer (PET) based detection of alkylating agents was reported. The detection is based on the quaternisation of a *Lewis* base nucleophile, usually a tertiary amine group that serves also as a quencher to the luminophore (De Silva *et al.*, 1997). Owing to the very simple structure of these alkylators, despite the high sensitivity attained by this approach still missing is the ability to identify the presence of specific alkylating agents, as with PET systems, most alkylators yield a very similar spectral response. This originates from the fact that the amine group that undergoes quaternisation is not π-conjugated to the luminophore and since in most cases the reaction of this amine with the alkylating agent yields very similar new C-N bonds. Therefore, in all cases the effect of the binding of the alkylator to the amine is expected to be very similar.

One approach to overcome this limitation is to electronically couple the nucleophile that reacts with the alkylating agent to the π-skeleton of the luminophore/chromophore so that its electronic properties are best affected by the nature of the alkylating group as well as its counter ion, if formed as a result of the alkylation process. Furthermore, as most alkylators are very similar in nature, the identification method should rely on reacting array of sensing spots with the same alkylator, seeking for specific reaction patterns that are unique to specific alkylators.

In the film, or when adsorbed onto high surface paper, materials **1-7** react at room temperature with vapors of various alkylating agents such as 1-chloro-2-ethylsulfanyl-ethane (a close analog of the sulfur mustard gas), **8**, chloromethoxy-ethane, **9**, iodomethane, **10**, and even dichloromethane, **11**, according to Fig. 2.

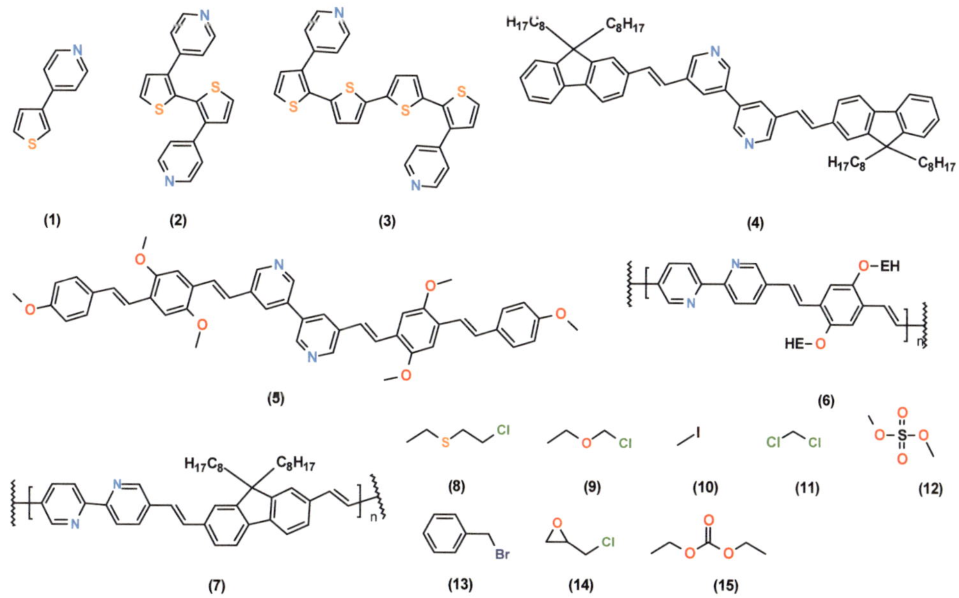

Fig. 2. Oligomers **1-7** and alkylating agents **8-15**

Fig. 3 shows the absorption and emission spectra of thin films of oligomer **5** and polymer **6** before and after alkylation with vapors of **8**. In both cases, the films of the nucleophiles readily react with vapors of the alkylator. Both the absorption and emission spectra of the nucleophiles are affected by the reaction with the alkylating agents. As in the product the alkylator is covalently bound to the film, the change in the optical characteristics does not revert when the concentration of the alkylator is lowered.

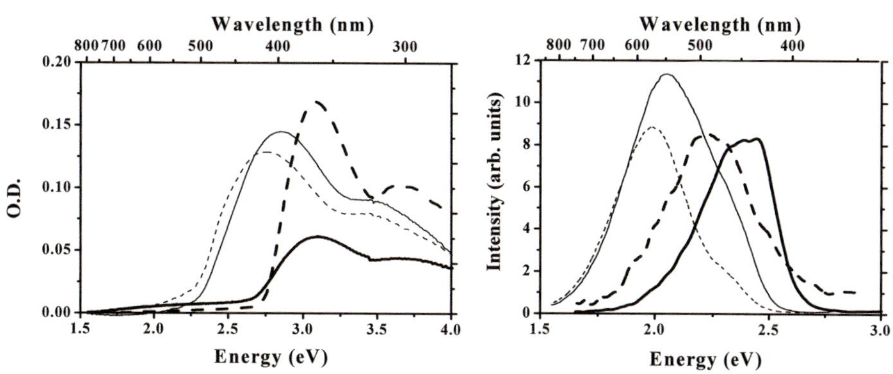

Fig. 3. Absorption (right) and emission (left) spectra of thin films of **5** (thick line) and **6** (thin line) on quartz before (continuous line) and after (dashed line) exposure to **8**

Inspired by the olfactory system (Dryer & Berghard, 1999) several groups developed artificial olfactory-like sensing arrays for detecting ions, volatile substances, amines, amino acids, proteins and carbohydrates where the identity of the target is defined by a response pattern of an array of low selectivity sensors. Similar approach was harnessed (Hertzog-Ronen *et al.*, 2009) for detection and identification of alkylating agents using a highly reactive and non-specific, multi-spot sensor system that relies on optical detection of luminescence spectral shifts for the identification step.

For that purpose, compounds **1-7** were adsorbed onto high surface filter paper squares (Whatman grade 50, low ash) and assembled into a seven-spot sensor array. Fig. 4 shows luminescence (λex=312 nm) pictures of the sensor array before (first row) and after (preceding rows) exposure to vapors of alkylating agents **8-15**. As can be seen from Fig. 4, one can clearly deduce the identity of the alkylating agent the sensor array was exposed to from the color pattern that is developed. For example, dichloromethane, **11**, and diethyl carbonate, **15**, produce a very similar response of luminescence shift upon reacting with **1**, **2**, **4**, **5**, **6** and **7**. With **3** the luminescence color shifts from RGB = [164, 84, 38] to [169, 135, 35] and [140, 152, 53], respectively. This colour difference is easily noticeable using a colorimeter and is even clear to the naked eye. Structurally similar **8** and **9** generate a very different luminescence shifts in all the spots.

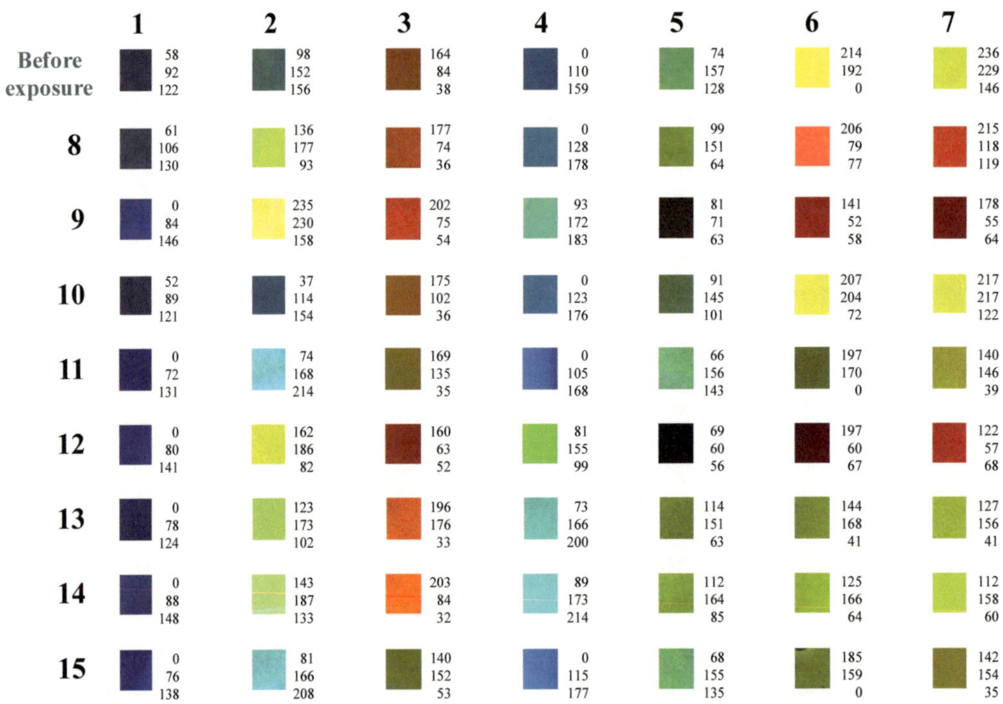

Fig. 4. A seven-spot sensor array of **1-7**, adsorbed on high surface filter paper, before (first row) and after (preceding rows) exposure to alkylating agents **8-15**. The figures on the right side of each square denote the RGB values of the spot when exposed to UV exciting light (λex=312 nm)

Principal component analysis (PCA) is a mathematical algorithm, useful for compression, classification and visualization of data. The purpose of this algorithm is to reduce redundant dimensionality of data by finding new trajectories, called principal components. These trajectories are orthogonal linear combinations of the original trajectories and along which the maximal variation in the data is retained.

The trajectories are chosen according to their descending sort order such that the first principal component (or trajectory) shows the largest variation and so on. The description along the new trajectories is more compact and can be referred as a shadow of the data when viewed from most informative viewpoint (Ringer, 2008). Datasets could be dramatically simplified using principal component analysis, Fig. 5.

A large scatter of the data is seen in the 3D plot of the delta RGB of the different alkylating agents with the different sensing spots of Fig. 4, Fig. 5 (left). This is expected to severely reduce the identification fidelity in real life applications, where data fluctuations originating from different sources are expected to contribute to the data scatter. To improve the identification ability of the system, PCA was applied to the data, Fig. 5 (right).

Not limited by association constant, this luminescence based sensing techniques for alkylating agents depend mainly on reaction rates. Such reactions can be easily accelerated by increasing mass transport, forcing the atmosphere to pass through or above the sensing unit, as well as by increasing the temperature of the sensing element. The sensitivity of the system depends solely on the number of reaction sites on the sensing element. This number is estimated to be $\approx 10^{14}/cm^2$, an equivalent of about 0.1 micrograms of mustard gas.

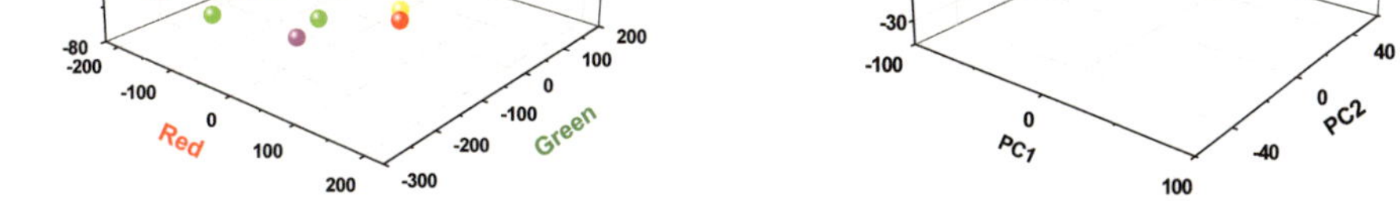

Fig. 5. (left) RGB plot for alkylating agent (8-20) and sensing spots (1-7) interaction; (right) Principal component analysis of 5 (left) data

2.2 Detection of alkylating agents using organic field effect transistors

Section 2.1 clearly exemplifies that monitoring the change in the optical signal of the molecule can be useful sensing tool for the detection of different alkylating agents. Still there is a need not only to sense the presence of a known material, but also to assess the toxicity of new molecules that produce unknown reaction pattern. Furthermore, machine reading of luminescence based detection signals require elaborate optical equipment, consisting of at least a light source and a multi-spot detector such as a CCD and a CPU. These two limitations motivated us to explore the possibility of direct electrical detection of alkylating agents.

The approach for direct electrical detection of alkylating agents relies on organic field effect transistors (OFETs) that carry an organic semiconductor that bears nucleophilic reaction sites and is thus capable of reacting with alkylating agents, see Fig. 6. The OFET in Fig. 6 consists of a heavily doped silicon base (a) that serves as the gate. This silicon layer is covered by a thin silicon dioxide layer that insulates the gate from the electrodes and semiconductor (b). The organic semiconductor is spread in between the drain and source electrodes, atop the silicon dioxide (c). The electrical detection process is based on the ability of the organic semiconductor to react with the alkylating agent (d) and change measurable quantities, such as V_{gate}, $V_{drain\text{-}source}$.

Direct electrical detection of chemical substances may be achieved in a variety of methods, of which field effect modulation of the conductivity of an organic layer seems most promising due to the potential of signal amplification and gain of information on the captured substance from different types of gate manipulations. The ability of building arrays of field effect transistors, each bearing a different active material offers promising routes to the detection and identification of diverse substances in solution as well as in the gas phase (Chang *et al.*, 2006).

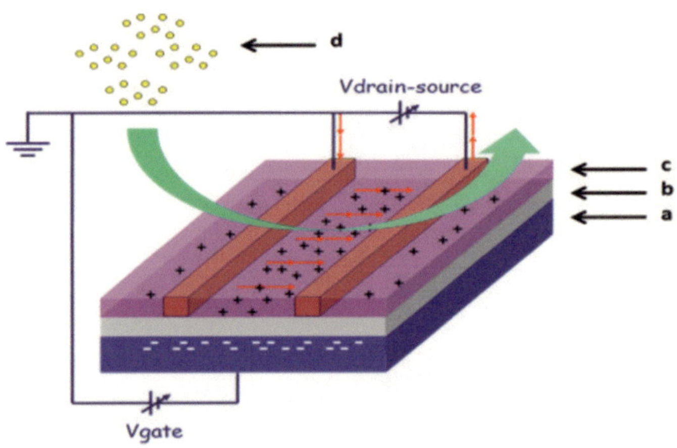

Fig. 6. Organic field effect transistor for detecting alkylating agents

Coupling of a nucleophile to an organic semiconductor was achieved by co polymerizing 4-[2,5-bis (2-thienyl)-3-thienyl] pyridine, **21**, with 3-hexylthiophene, **22**, to obtain co-polymer **23**, as shown in Fig. 7.

Co-polymer **23** was obtained as a bright red solid, Mn = 37000, Mw = 97000 (GPC), in 17% yield by co-polymerization of a 1:10 solution of **21** and **22** in chloroform in the presence of $FeCl_3$. NMR characterization reveals that co-polymer **23** is composed of $\mathbf{21}_n\mathbf{22}_m$ where n = 0.95±0.005 and m = 0.05±0.005 (Gannot *et al.*, 2010). Co-polymer **23** reacts with alkylating agents according to Fig.8, both in solution and as a film when treated with their vapors.

Co-polymer **23** is an excellent film forming polythiophene organic semiconductor. Bottom contact organic field effect transistors (OFETs) were fabricated by spin-coating a 50 nm film of the co-polymer atop the FET structure using a previously described procedure (Shaked *et al.*, 2003). Fig. 9 shows the electrical characteristics of the OFET having co-polymer **23** as the channel. Fig. 9 (left) shows the drain-source currents of the OFET as a function of drain-source voltage for several gate-source voltages (V_{GS} = 0, 10 and 20 V) before (black lines and inset) and after (gray lines) exposure to the alkylator methyl iodide (CH_3I, **10**). Fig. 9 (right) shows the drain-source current as a function of gate-source voltage for drain-source voltage of V_{DS} = 10 V measured the same OFET having co-polymer **23** as the channel and under the same conditions.

As can be clearly seen in the figure, prior to exposure to the alkylating agent the device exhibits standard FET output characteristics with a slight indication of doping, probably due to the exposure to atmospheric oxygen. The post exposure curves show considerably higher currents and a resistor like response while the effect of the gate bias is lost. This is a clear indication for high level p-type doping associated with the exposure to methyl iodide.

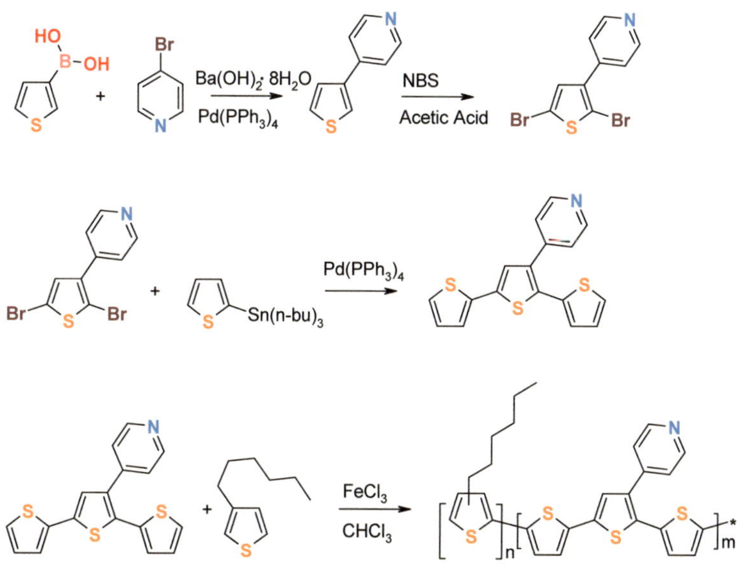

(21) (22) (23)

Fig. 7. Preparation of co-polymer **23**

(23) (23-R)

Fig. 8. Reaction between co-polymer **23** and an alkylating agent

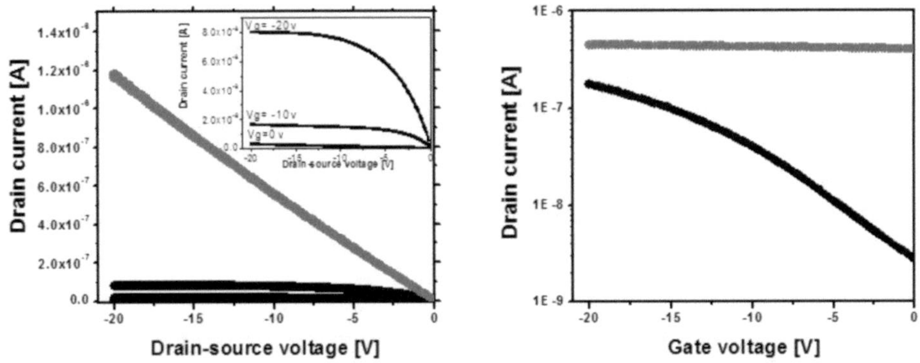

Fig. 9. Output (left) and transfer (right) characteristics measured on a FET having co-polymer **23** as the channel. The black lines and the inset show the device response prior to exposure while the gray (overlapping in a.) lines show the response after exposure to methyl iodide, **10**.

Consistent with the results presented in Fig. 9 (left), the transfer characteristics of the OFET prior to exposure show current modulation due to the gate-source voltage. The post exposure gray horizontal line shows almost no dependence on the gate voltage. At zero gate bias, $V_{GS} = 0$, the exposure of the OFET to methyl iodide enhances the current by almost two orders of magnitude while at V_{GS} = -20 V the factor is only 2-3.

The detection of alkylating agents using co-polymer **23** OFET, involves the formation of a new covalent bond between the alkylating agent and the co-polymer in the channel. As this process is not reversible, all available reaction sites are bound to react. The number of reaction sites (pyridine rings, see Fig. 8) represents also the upper detection limit of the device. The total weight of the polymer embraced by the source and drain electrodes, M, is 10^{-9} g. (assuming a film density of 1g/cm³), Equation 1, where W is the width of the device, L is the length and h is the thickness of the polymer film.

$$M = W \times L \times h \times p = 1 \times 2{\cdot}10^{-4} \times 5{\cdot}10^{-6} \cdot 1 = 10^{-9}\ \text{g} \qquad (1)$$

Co-polymer **23** consists of 5% (5×10^{-11}g) monomer **21,** which are equivalent to ~1.5×10^{-13} moles. Consequently, reacting the OFET with ~9×10^{10} analyte molecules will bring the system to saturation of all the active sites of the polymer.

The OFET of co-polymer **23** produced similar results when exposed to a variety of alkylating agents, such as chloromethoxy ethane, **9**, 1-chloro-2-ethylsulfanyl ethane (a close analog of sulfur mustard gas), **8**, and benzyl bromide, **13**. In contrast, exposure of the OFET to vapors of non reactive materials, (see Fig. 10), that are incapable of forming covalent bonds with co-polymer **23**, such as methanol, **24**, acetone, **25**, water, **26**, and isopropanol, **27**, resulted with the deterioration of the output and transfer characteristics, as can be seen in Fig. 11.

The effect of alkylation on OFETs having co-polymer **3** as the channel can be rationalised by following the effect of alkylation on the charge distribution in the polymer. Co-polymer **23** is composed of electron rich polythiophene skeleton with pendant pyridine groups. Upon reacting with an alkylating agent, the pyridine group is transformed into an electron deficient pyridinium group. These electron poor groups attract charge density from the polythiophene skeleton, thus partially doping it.

As indicated above, the OFET readily reacts with any alkylating agent, producing the same type of reaction and thus cannot provide any straightforward information regarding the identity of the reacting alkylator. The OFET can provide valuable information regarding the reactivity of the alkylator. Fig. 12 shows the drain-source current as a function of the exposure time to vapors of methyl iodide, **10**, at a constant gate-source voltage of -10 V. When methyl iodide vapors are introduced into the atmosphere surrounding the OFET (marked with an arrow) the drain-source current drops, indicating the process of absorption of the vapors by the polymer and associated changes in its morphology and packing. Unlike "innocent" materials, that are incapable of reacting with the organic semiconductor, alkylating agents react in a way that increases the source-drain current at a given V_{gate}. This offers a simple and effective way of assessing the reactivity of the alkylating agent as the initial drop in source-drain current is a simple measure of the concentration of the guest in the semiconductor host while the rate of current increase at a fixed gate bias is a measure of the rate of alkylation.

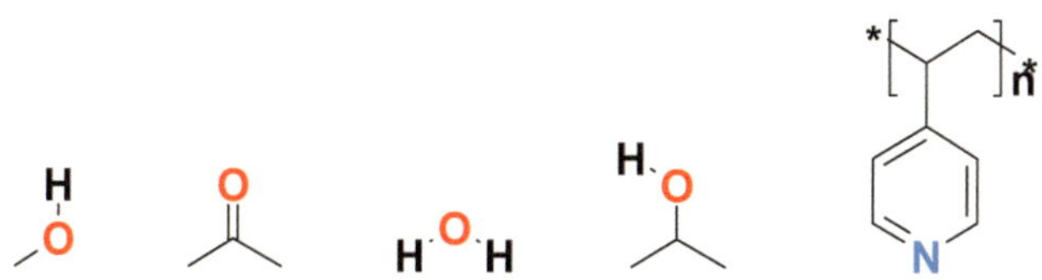

(24) (25) (26) (27) (28)

Fig. 10. Structures of non reactive materials, **24-28**

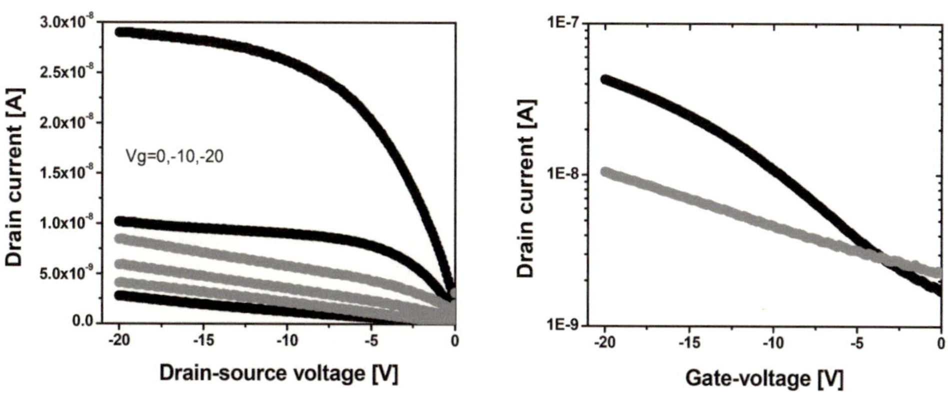

Fig. 11. Output (a) and transfer (b) characteristics of OFET having co-polymer **23** as the channel. The black lines show the device response at ambient conditions, while the gray lines show the response after exposure to methanol, **24**.

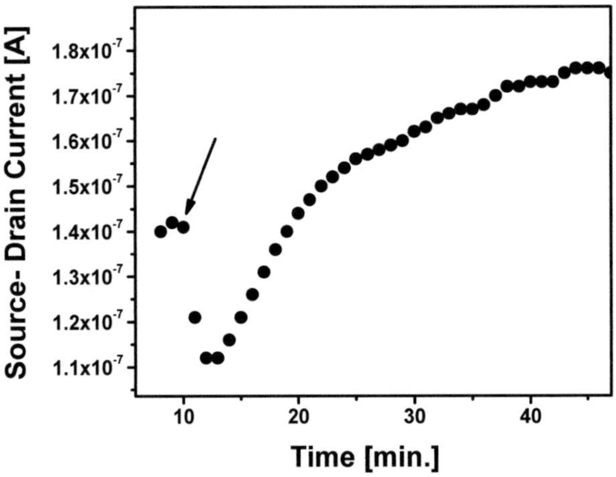

Fig. 12. Time response of an OFET having co-polymer **23** as the channel. Methyl iodide, **10**, is introduced at T=10min (arrow).

2.3 Detection of alkylating agents using chemo-mechanical asymmetric micro-cantilever devices

Alkylation of nucleophilic species is capable of inducing irreversible effects other than changes in electrical and optical signals. For example, the alkylation of a polyamine polymer may induce a volume change. This, in turn, may be translated into a macroscopic mechanical motion. The possibility of harnessing a chemo-mechanical signal transduction technique on micron-size micro-cantilevers for the detection of alkylating agents was explored.

Micro-cantilevers have gained much attention in recent years as a promising platform for biomechanical as well as chemo-mechanical sensors (Lavrik *et al.*, 2004). Different groups have reported the ability to detect different chemical substances, including explosives and warfare agents (Stolyarova *et al.*, 2010). Micro-cantilever based systems also allow label-free (Raorane *et al.*, 2008) real time monitoring of chemical and biological processes such as sequence specific DNA hybridization (Wu *et al.*, 2001) and drug-target binding interactions (Ndieyira *et al.*, 2008). Recently, due to their high sensitivity, micro-cantilevers have been used to measure the physical properties of polymers near their glass transition temperature (Jung et al., 2008). Typically, micro-cantilevers can be operated both in static and dynamic modes. In the static mode, the bending (deflection) of the tip serves as a measure for the interaction between the micro-cantilever and the guest while in the dynamic mode the shifts in the resonance peak indicate the change in the mass and stiffness of the micro-cantilever. In a static bending mode it is essential to asymmetrically functionalize the micro-cantilever so a stress gradient across it can be developed and induce bending. Novel asymmetric porous-silicon-over-silicon (PSOS) micro-cantilevers made according to Fig. 13, using vapour phase stain etching (Stolyarova *et al.*, 2008), offer inherently asymmetric, highly sensitive and easy to coat systems. The resulting micro-cantilevers are characterised by a rough surface on their lower side and an intact flat surface on their upper side, Fig. 13 (b) and (c), respectively.

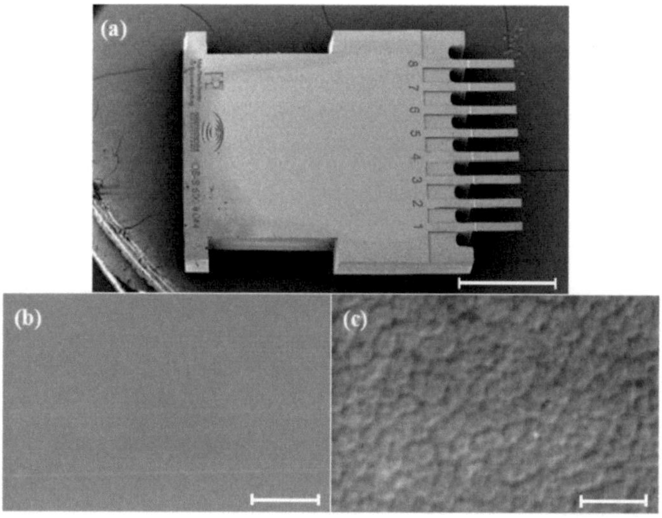

Fig. 13. (a) A SEM micrograph of a microcantilever array used in this study as the transducer. Representative SEM images of the upper (masked, b) and lower (unmasked, c) surfaces of the polymer coated microcantilever depicted in a. Bar size: 1 mm in a) and 5μm in (b) and (c).

Unlike other methods of polymer application including drip coating and evaporation, the PSOS microcantilevers offer a simple way of coating. Due to their inherent asymmetric surfaces, a coating polymer layer may be applied to the beams by simple dipping in a polymer solution, rinsing and drying. For the detection of alkylating agents, micro-cantilevers were coated with poly-4-vinylpyridine, **28**, serving as the sensing layer. Fig. 14 shows a side view of the polymer-coated micro-cantilever chip under gas flow. As can be seen, the dominant polymer layer resides on the lower surface of the micro-cantilever. This surface is characterised by rough porous silicon layer. The micro-cantilever chip was mounted in a flow cell where both the temperature and the flow were controlled. The bending of the micro-cantilever tip was monitored optically using optical profiling system capable of running time-resolved measurements. To monitor the effect of different analytes on the bending pattern, nitrogen served as an inert carrier gas and a special evaporation chamber allowed mixing the analytes with nitrogen. During baseline and desorption stages the polymer microcantilever chip was exposed to pure nitrogen, while during absorption to a mixture of nitrogen and one of the analytes (Shemesh *et al.*, 2011).

Fig. 15 shows the bending patterns of the micro-cantilever tip at absorption and desorption stages for two different analytes.The first one is vapors of methanol, **11**, and then vapors of methyl iodide, **10**, both in nitrogen carrier gas. As can be clearly seen, the absorption of coated microcantilever is fully reversible as ethanol is bound to the polymer only through weak and labile bonds. Concequently, ethanol readily desorbs from the micro-cantilever under an ethanol-poor atmosphere. The result of the desorption is that the bending of the micro-cantilever returns to its initial level, as it was before adsorption of the ethanol. In contrast, upon exposure of the same micro-cantilever to methyl iodide, **10**, reaction occurs between the polymer and the analyte. This reaction attaches the analyte to the polymer through a new and stable covalent bond. Since this reaction is not reverted at room temperature, the bending is irreversible. This is reflected during the desorption stage as nitrogen flushes the system. The bending of the micro-cantilever reaches new bending level closer to the adsorption steady state level, indicating that some unreacted methyl iodide desorbs from the surface of the micro-cantilever but some of the methyl iodide had reacted with the polymer and cannot desorb.

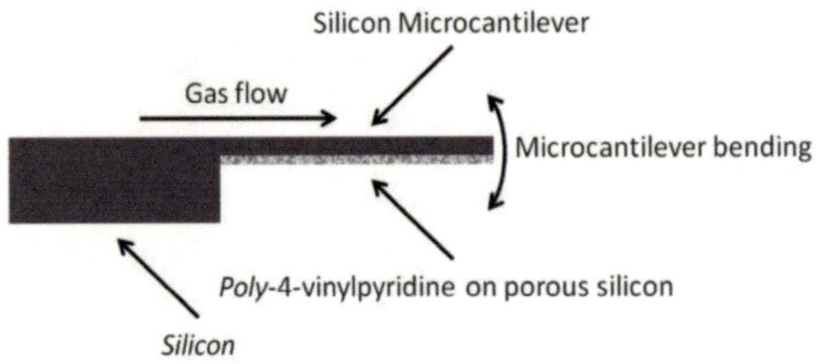

Fig. 14. Side view of the polymer coated microcantilever chip.

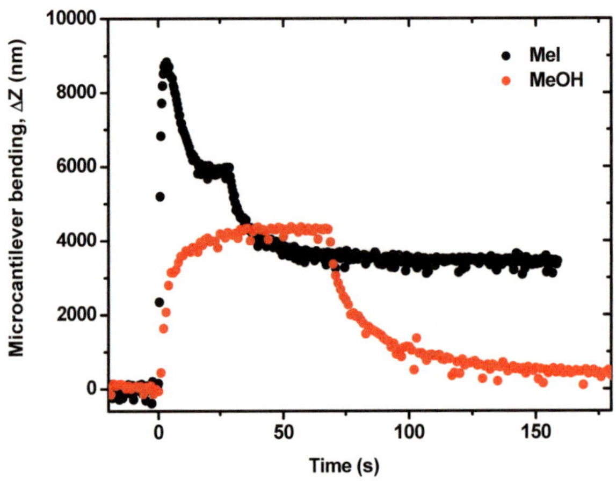

Fig. 15. Micro-cantilever bending curves for methanol, **11**, and methyl iodide, **10**,

3. Conclusions

Luminescent, electrical and chemo-mechanical detection schemes for the sensing (detection, identification and risk assessment) of alkylating agents were explored. A multiple-spot "chemical nose" array is shown to offer both detection and identification of different alkylating agents in the gas phase. The rate of colour development offers some risk assessment ability (not discussed here). Since the sensing element is based on a chemical reaction the sensitivity of the sensor is very high and depends mainly on mass transport of the alkylating agents and temperature.

In a similar way and based on very similar nucleophile-electrophile chemistry organic field effect transistors and micro-cantilevers were harnessed to the detection of alkylating agents.

Reaction of gas phase alkylators with a nucleophilic organic semiconductor in an OFET structure yields dramatic and irreversible changes in the transistor properties while other innocent molecules inflict reversible changes to the OFET parameters. Risk assessment was demonstrated by following the absorption of the analyte and its reaction with the organic semiconductor.

In a similar manner, gas phase alkylators that react with a nucleophilic polymer at the surface of a micro-cantilever induces its irreversible bending while for innocent materials, such as alcohols, the bending is reversible and depends on the momentary concentration of the guest in the atmosphere.

4. Acknowledgements

The research leading to these results received funding from the European Union Seventh Framework Programme under grant agreement No :2015180 [COMMONSENCE] and from the Technion Funds for Security Research.

5. References

Borzin, E.; Shemesh, A., Hertzog-Ronen, C., Gerchikov, Y., Tessler, N. & Eichen, Y. (2010). Towards an Omnipotent "Artificial Nose", Detection and Identification of Alkylating Agents Using an Optical Sensor Array. *Journal of Physical Organic Chemistry*, vol. 23, pp. 1108-1113

Chang, J.B.; Liu, V., Subramanian, V., Sivula, K., Luscombe, C., Murphy, A., Liu, J. S. & Frechet J. M. J. (2006). Printable Polythiophene Gas Sensor Array for Low-Cost Electronic Noses. *Journal of Applied Physics*, vol. 100, pp. 014506

De Silva, A.P.; Gunaratne, H. Q. N., Gunnlaugsson, T., Huxley, A. J. M., McCoy, C. P., Rademacher, J. T. & Rice, T. E. (1997). Signaling Recognition Events with Fluorescent Sensors and Switches. *Chemical Reviews*, vol. 97, pp. 1515-1556.

Dryer, L. & Berghard, A. (1999). Odorant Receptors: a Plethora of G-Protein Coupled Receptors. *Trends in Pharmacological Sciences*, vol. 20, pp. 413-417.

Gannot, Y.; Hertzog-Ronen, C., Tessler, N. & Eichen, Y. (2010). Sensing of Alkylating Agents Using Organic Field-Effect Transistors. *Advanced Functional Materials*, vol. 20, pp. 105-110

Hertzog-Ronen, C.; Borzin, E., Gerchikov, Y., Tessler, N. & Eichen, Y. (2009). Detection and Identification of Alkylating Agents Using a Bio-Inspired "Chemical Nose". *Chemistry - A European Journal*, vol. 15, pp. 10380-10386.

Jung, N.; Seo, H., Lee, D., Ryo, C. Y. & Jeon, S. (2008). Nanomechanical Thermal Analysis of the Glass Transition of Polystyrene Using Silicon Cantilevers, *Macromolecules*, vol. *41*, pp. 6873-6875

Lavrik, N. V.; Sepaniak, M. J. & Datskos, P. G. (2004). Cantilever Transducers as a Platform for Chemical and Biological Sensors. *Review Scientific Instruments*, vol. *75*, pp. 2229-2253.

Ndieyira, J. W.; Watari, M., Barrera, A. D., Zhou, D., Vogtli, M., Batchelor, M., Cooper, M. A., Strunz, T., Horton, M. A., Abell, C., Rayment, T., Aeppli, G. & Mckendry, R. A. (2008). Nanomechanical Detection of Antibiotic-Mucopeptide Binding in a Model for Superbug Drug Resistance. *Nature Nanotechnology*, vol. *3*, pp. 691-696.

Raorane, D. A.; Lim, M. D., Chen, F. F., Craik, C. S. & Majumdar, A. (2008). Quantitative and Label-Free Technique for Measuring Protease Activity and Inhibition using a Microfluidic Cantilever Array. *Nano letters,* vol. *8*, pp. 2968-2974.

Ringner, M. (2008). What is Principle Component Analysis. Nature Biotechnology, vol. 26, pp. 303-304.

Shaked, S.; Tal, S., Roichman, Y., Razin, A., Xiao, S., Eichen, Y. & Tessler, N. (2003). Charge Density and Film Morphology Dependence of Charge Mobility in Polymer Field Effect Transistors. *Advanced Materials*, vol. 15, pp. 913-916.

Shemesh, A.; Stolyarova, S., Nemirovsky, Y. & Eichen, Y. (2011). Isotope Effect in the Interaction between Gas-Phase Isotopologues and Polymer-Coated Porous Silicon Over Silicon Microcantilevers. *The Journal of Physical Chemistry C*, vol. 115, pp. 15980-15987.

Stolyarova, S.; Cherian, S.; Raiteri, R.; Zeravik, J.; Skladal, P.; Nemirovsky, Y. (2008). Composite porous silicon-crystalline silicon cantilevers for enhanced biosensing, *Sensors and Actuators B*, vol. *131*, pp. 509-515.

Stolyarova, S.; Shemesh, A., Aharon, O., Cohen, O., Gal, L., Eichen, Y., & Nemirovsky, Y. (2010). Vertically Integrated MEMS SOI Composite Porous Silicon- Crystalline Silicon Cantilever-Array Sensors: Concept for Continuous Sensing of Explosives and Warfare Agents. *NATO Science for Peace and Security Series B: Physics and Biophysics*, pp. 261-274.

Tal, S.; Salman, H., Abraham, Y., Botoshansky, M. & Eichen Y. (2006). Sensitive and Selective Photoinduced-Electron-Transfer-Based Sensing of Alkylating Agents. *Chemistry European Journal*, vol. 12, pp. 4858-4864.

Wu, G.; Ji, H., Hansen, K., Thundat, T., Datar, R., Cote, R., Hagan, M. F., Chakraborty, A. K., & A. Majumdar (2001). Origin of Nanomechanical Cantilever Motion Generated from Biomolecular Interactions. Proceedings of the National Academy of Sciences of the United States of America, vol. 98, pp. 1560-1564.

6

Gas Sensor Array with Broad Applicability

Andrzej Szczurek and Monika Maciejewska
Wrocław University of Technology
Poland

1. Introduction

In recent years, effort has been made to develop instruments for rapid, inexpensive analysis of volatile chemical species that do not require trained personnel. This demand has been mainly driven by a variety of real life applications. Indeed, the problem of classifying and further quantifying chemical substances on a real-time basis is very critical for a broad range of activities in various fields, like: industrial (Garrigues *et al.*,2004), agricultural (Berna, 2010), medical (Byun *et al.*, 2010), domestic (Zampolli *et al.*, 2004) and environmental (Bourgeois *et al.*, 2003).

Rapid detection and quantification of chemical species are important in control and optimisation of industrial processes and bioreactors. Optimisation of combustion processes can lead to significant energy savings, as well as to minimization of emissions across power, chemical, steel, paper, food and other manufacturing industries. The ability to monitor and precisely measure leakages of combustible and explosive gases is crucial in preventing the occurrence of accidental explosions. Nowadays, a lot of attention is paid to safety monitoring, surveillance and homeland security. Quality control of different products, e.g. food, drinks and agricultural products, are also on the rise.

New demand has recently emerged in the area of medical diagnosis and continuous monitoring of chemical parameters during intensive care. Medical appliances are related mainly to the breath analysis and detection of infections, diseases and bacteria. Environmental applications include e.g. monitoring of air quality (indoor and outdoor), potentially polluting industrial installations, plants and soil contaminations, landfill sites, wastewater treatment plants.

Rapid advancements in sensing technology result in a variety of devices which are able to detect indoor pollutants. They have found widespread commercial application in gas monitoring and alarm applications. New trends in building engineering cause that a distributed control of ventilation is required. Three factors are considered in this kind of monitoring: comfort, productivity, security (agents, pollutants, explosives).

In the past decade there has been a growing interest in the development of olfactory machines and electronic nose systems. Odour monitoring in ambient air is a serious challenge for the measurement technology. There is no technique available for continuous measurement of odour nuisances and the current standard methods of odour measurements cannot be applied to on-line monitoring.

In the applications mentioned above, devices based on gas sensors are especially applicable. They have several advantages over conventional analysers, because of possible miniaturization, low price and maintenance costs, short response time, easy manufacturing and small size. Currently, gas sensors play important, commercial role in detecting, monitoring and controlling the presence of hazardous and poisonous gases in the atmosphere.

Gas sensors are one kind of chemical sensors. According to the current IUPAC's definition, a chemical sensor is a *device that transforms chemical information, ranging from the concentration of a specific sample component to total composition analysis, into an analytically useful signal*. The chemical information may originate from a chemical reaction of the analyte or from a physical property of the system investigated. All chemical sensors consist of a chemically sensitive material (a receptor) that is interfaced to a transducer. In the first unit the chemical information is transformed into a form of energy, which may be measured by the transducer. The receptor is responsible for selectivity of measurements. The transducer part is a device capable of transforming the energy which carries the chemical information about the sample into a useful analytical signal. The transducer as such does not show chemical selectivity. The interaction between an analyte molecule and the receptor can be either reversible (measurand dissociates from the layer when the external concentration is removed) or irreversible (measurand undergoes a chemical reaction and the sensitive material layer is consumed).

The receptor and transducer parts of chemical sensors may be based on various principles. Therefore, these devices have different application ranges. They can be used for:

- detection of some of chemical conditions, e.g. the presence or existence of strictly defined chemical, harmful hazards;
- specific determination of one substance;
- classification of complex gases (determination of classes);
- qualitative and quantitative chemical analysis of multi-component gaseous mixtures.

The application range of gas sensors is defined by their measurement characteristics. The traditional strategy in the sensor techniques is based on a single sensor to convert chemical information to an electric signal. A principle of "*lock-and-key*" is applied in the design of this device. Selectivity is achieved through recognition of the analyte molecules at the receptor site. In this approach, specific receptor has to be synthesized to bind the analyte of interest strongly and highly selectively. Therefore, the traditional strategy requires highly selective sensor for each analyte under test. This approach is not particularly useful for at least two reasons. Firstly, the synthesis of the separate, specific receptor is not easy to realize. It is a challenge both from chemical and technological point of view. Often, this requirement cannot be achieved due to the working mechanism of the sensor, especially its receptor and transducer functions. Secondly, this approach is impractical for analysing complex vapour mixtures qualitatively and quantitatively. In practice, receptors do not ensure very good selectivity. However, in certain circumstances, under carefully controlled operating conditions, the output signal may provide selective information about individual component, without any major preliminary treatment of the sample. Therefore, conventional sensors are normally designed to operate under well defined conditions, for specified analytes, in certain sample types. They offer rapid and inexpensive detection of many individual volatile substances with simple interpretation of measurement results.

The ideal gas sensor should exhibit reliability, robustness, sensitivity, selectivity and reversibility. These requirements are difficult to attain in practice. For example, often sensing elements cannot achieve the required selectivity. It is a serious problem, because the key function of the analytical equipment is to realize a kind of selectivity for the quantity to be measured. One of the main objectives of current research in the gas sensor technology is the qualitative and quantitative multi-component analysis of chemical environment characterized by the simultaneous occurrence of different volatile species in air. Sensitivity limitation originates from fundamental principles of the sensing mechanism, which is different from "*lock-and-key*" principle. This disadvantage cannot be eliminated completely because of sensing mechanism of gas sensors. However, there are methods to reduce the influence of interfering chemicals. Recently, there has appeared a strategy that is complementary to traditional chemical sensing. It involves using systems based on:

- sensor array;
- signal processing;
- data analysis.

The advantage of this strategy consists in application of unspecific but satisfactorily reproducible sensors for multi-component analysis of gases.

2. Sensor array

Generally speaking, a sensor array is a set of sensors used for gathering information about the object under test. In chemical applications, it consists of several different sensors with broad and partially overlapping sensitivity to various gases. The gas sensor array is used to convert chemical information regarding multi-component gaseous mixtures into a set of measurable signals. Sensors are exploited independently and simultaneously in this device. Therefore, they can be treated as the independent sensing elements in the operation procedure. The multivariate response of a sensor array is created by all sensing elements. This collection of sensors should be characterised by as much chemical diversity as possible. In this case, the array responds to the largest possible range of analytes.

The first report of the sensor array was presented by Persaud and Dodd in the early 1980s (Persaud & Dodd, 1982).The utilization of these devices is inspired by the performance of biological olfactory systems. The sensor array is established on an assumption that a cross-sensitivity of gas sensors is unavoidable. The cross sensitivity means that some chemicals may interact to give a different signal from the component in a mixture compared to the single component. For that reason, instead of trying to eliminate this feature, gas sensors are linked as independent sensing elements in an array configuration. Of course, the principle of "*lock-and-key*" is abandoned in this approach. Sensing elements are not highly selective toward any given analyte. However, they should be satisfactory reproducible and they should have significantly different gas sensing properties. The selectivity of each sensing element is admittedly low. However, the combination of the responses of different sensors presents a characteristic pattern that can be treated as a unique 'signature' ("electronic fingerprint") of individual chemical species. Subsequent signal processing and data analysis are required to extract information about gas under examination.

A sensor array can provide both qualitative and quantitative information. It shows the ability to classify different complex samples and to quantify components concentrations when the mixture consists of several constituents. In other words, the array of sensors performs integration to yield a unique signal for complex but distinctive gaseous samples. It is realised without requiring that the mixture is broken down into its individual components prior to, or during, the analysis. This is a disadvantage when the precise qualitative and quantitative information about a complex mixture is necessary, but it is advantageous when only classification of the sample is required. Another potential shortcoming to an array system is the possibility that other unknowns may give the same "unique" signal as a specific analyte of interest. Qualitative and quantitative analysis can be performed only in case the sample consists of several components.

3. Design of gas sensor array

A gas sensor array never functions as an independent analytical instrument. It is one of the few essential components of a detector, an analyser, a monitor (the analyser working according to a sampling plan as a function of time) or an automated system for continuous measurements, in real time. For that reason, the sensor array must cooperate with other elements of the analytical equipment. Usually, the following functions are performed by external settings :

1. sampling,
2. transport and handling of the sample,
3. signal acquisition and processing,
4. data analysis,
5. reporting and visualization of measurement results,
6. automatic control of the whole system,
7. calibration.

The sensor array construction is particularly dependent on the first four functions. Therefore, the following issues are significant for sensor array design:

- method of sampling, pre-treatment and target gas delivery equipment;
- measurement characteristics of a sensor array;
- working parameters;
- operation mode;
- signal processing;
- data analysis.

Additionally, usability, service and cost of sensor array should be taken into account at the design stage. Improvement in any of these areas will lead to a significantly better performance of the analytical equipment.

3.1 Method of sampling, pre-treatment and gas delivery equipment

A sensor array may be adjusted for dynamic (active) or passive (diffusion) sampling. In the first case, the sample is mechanically drawn into the sensor array, where it is analysed, and then exhausted to the atmosphere or the vent line. This type of sampling is performed by a suitable pneumatic system. This assembly consists of a sample probe and a delivery system, which is designed to transfer the gas from the source to the sensor array. Usually, the delivery system includes a gas line, an electro-valves, an electronic flow meter (a mass flow controller), a gas mover (e.g. a pump or a fan). The intrinsically safe or explosion-proof pump has to be employed in some applications. The sampling system must be designed to draw continually a fresh sample of the target gas. To ensure the unaltered state of the sample:

- the sampling probe and the gas line must be made of inert materials such as: polytetrafluoroethylene (PTFE), tetrafluoroethylene hexafluoropropylene copolymer (FEP), polyethylene terephalate (PET), glass, stainless steel;
- the transport line should be short;
- particulate matter and moisture cannot accumulate inside the sampling line.

A sample probe and gas line has to be cleaned between successive exposures until all contaminations are removed. Usually, a stream of pure air is used for this operation. Thus, a sensor array has to be connected with a source of this gas. In the dynamic sampling, a sample can be taken directly from a surrounding atmosphere, an industrial installation as well as from a headspace sampler, a bubbler or a pre-concentrator.

A passive (diffusion) sampling operates by allowing gas or vapour molecules to diffuse until they reach sensor surface. Fick's First Law describes the movement of gas. As the diffusion coefficient varies in a known manner with temperature and pressure, these parameters must be taken into account while designing the sensor array. In this kind of sampling, sensor array is often enclosed in explosion-proof housings (if it is required by the conditions in the place of installation), and signals are delivered back to the control panel. The enclosure is described as a "sensor head". The shape and construction of this package affect the measurement characteristics of the sensor array.

In diffusion sampling, the delivery system is not exploited. This is an important advantage in certain applications. For example, as a pump is not used to move the sample, the analytical equipment is completely quiet. Sampling based on diffusion relies on air movement rather than on actively pulling a sample. Therefore, the response time of an array is usually slower in this case. When diffusion sampling is applied calibration of the sensors is often complicated. There are required special accessories to convert a diffusion operation mode to flow-through for calibration purposes. Moreover, the implied equivalence between calibrating via flow-through and monitoring under diffusion is not always well-documented. Furthermore, all gas sensors measure partial pressure, and a sample actively brought to the sensor is at a slightly elevated pressure, while a diffusion sensor operates at ambient pressure. As such, the output sensitivity of sample draw sensors is usually higher than diffusion sensors. This can be important for many toxic gases with low regulatory levels.

In many applications, it may be necessary to pre-treat sample before it reaches a chemical sensitive layer of a sensor. For example, filtration, moisture removal or chemical modification may be required. Different modifications of a delivery system are discussed in the literature (Stetter *et al.*, 2000). For example, the system may be equipped with molecular filters (like charcoal) or a heated filament. If air is used as a carrier gas, the filament performs catalytic oxidation. In case the volatile compounds are in an inert gas (without oxygen), the term pyrolyzer is more appropriate for the filament. A temperature- variable filament is particularly interesting because it provides additional selectivity at the expense of time. As it was mentioned, sensors demonstrate poor selectivity in real analytical applications. Chromatographic columns can be used to separate constituents of the complex mixtures on the basis of their molecular size or other physical properties prior to analysis. A sensor array can also be connected to a membrane unit. Its material is usually used for separation, purification, and chemical enrichment of the sample. These functions are very well performed by Nafion. This material is a flouropolymer with ion exchange capacity. It belongs to the class of solid polymer electrolytes. Nafion has a hydrophobic $-CF_2-CF_2-$ and hydrophilic $-SO_3H$ regions in its polymeric structure. Due to these properties Nafion has high permeability for water. Thus this material can be used as a membrane drier. Nafion also removes other compounds from the sample stream e.g. volatile alcohols (methanol, ethanol), acetone and some other polar solvents. This material is really a separator and it fractionates the sample in a predictable and repeatable manner prior to analysis by the chemical sensor array. Therefore, Nafion is capable of enriching or depleting the gas stream with any number of analytes, to which it is permeable.

3.2 Measurement characteristics of gas sensor array

Usually, the responses of individual sensors (especially commercial ones) to various substances are only slightly different. Therefore, various methods have to be used to achieve satisfied measurement characteristics of the sensor array. They are based on:

- dimension and composition of a sensor array;
- construction of this device;
- properties of the individual sensors.

3.2.1 Dimension and composition of gas sensor array

Dimension of a sensor array is defined by the number of sensors. This factor strongly affects measurement characteristics of the sensor array (Gardner *et al.*, 2005; Gualdr´on *et al.*, 2006). Large number of sensors may in some situations improve discrimination abilities of the sensor array. However, a calibration effort and time demand for data processing increase considerably in this case. Additionally, multi-dimensional arrays may include sensors, which are not sensitive to target gases and do not contribute to the recognition task. Sensors with no responses to the subgroup of analytes increase noise and degrade the ability to make analytical comparisons. Besides irrelevant devices, an array can consist of sensing elements which have very similar sensitivities to tested volatile compounds. The redundant sensors, which are exactly the same, lead to improvements in sensitivity. Perfectly redundant sensors increase the signal-to-noise ratio (SNR) and produce a better low detection limit (LDL). However, these devices provide redundant information which is not useful for the discrimination process. The removal of redundant sensors offers many advantages, e.g. level of drift/noise introduced by these devices can be greatly reduced. Moreover, the description of sensor array response is easier. Thus, simpler classifiers (working on small number of dimensions) can be applied and more qualitative information is obtained. Therefore, arrays with smaller number of sensing elements, but with different response parameters are preferred in practice. By eliminating unimportant sensors, the cost and time of collecting and analysing data may be reduced. This is crucial for many applications. Of course, the range of applications of the sensor arrays with the small number of sensing elements is decreased.

It is important to select sensors not only from the perspective of sensor array dimension. This device may include sensors of different classes, which have different properties. Therefore, sensor arrays are divided to two groups: homogeneous and heterogeneous (Tomchenko *et al.*, 2003). For example, the earliest reported electronic nose was based on a heterogeneous array of combustible and electro-chemical sensors. It was reported in (Stetter *et al.*, 2000) that sensors with chemically independent responses are valuable and make the array more versatile and able to distinguish more analyte differences. Sensors of various classes generally provide more chemically different or chemically independent responses than sensors of the same class with small variations in their formula or structure. Sensors, whose chemical principles are different, give data that are more effective when comparing samples than the same number of sensors of a single class. Moreover, in large arrays of different sensors of one class, sensors that may not contribute with information will always contribute with noise. There is an optimum array composition for any given set of analytes in qualitative detection applications.

There are two possible approaches to design sensor arrays. The first one is commonly applied in commercial instruments e.g., in electronic noses. This strategy is based on an assumption that sensor array should be dedicated to a limited number of chemical species emitted by e.g. food or other goods, human organisms, industrial or agricultural installations. To realize this goal, the set of suitable gas sensors is chosen. The selection process is related to the particular applications and it is performed at the stage of instrument design. The inadequate choice of sensors can potentially result in insufficient information or redundancy. Therefore, the optimisation procedure has to be applied. The considerable efforts have been made in this sector of sensor technique recently. However, the scope of application of sensor arrays constructed in this way is rather limited.

The alternative approach concerns sensor arrays of more universal applicability. These devices require using a broad spectrum of sensing elements, so that a large number of gases could be measured by one analyser or analytical system. Due to progress in the sensor technology the construction of relatively cheap arrays consisting of many sensing elements is already possible. Of course, the problem of sensor selection exists in this strategy too. However, it is solved in another way. We propose to use in the analytical equipment an array with a large number of sensors. They are not selected with a particular application in mind. However, they shall cover as broad range of substances as possible in a partially selective manner. For particular applications from this pool of the sensing elements there are chosen various sensor combinations with conscious intent towards the analytical goal. These sensor subsets have lower dimension than the original sensor array. Additionally, they have optimal measurement characteristics.

3.2.2 Construction of sensor array

Many contributions can be found in literature devoted to the characterization of individual gas sensors. However, little attention has been paid to the construction of a sensor array. This approach is inadequate because construction of this device decides about:

- chemical environment of the sensors;
- space distribution of the examined gaseous species inside measurement chamber and its evolution in time;
- thermal, electromagnetic or electrical conditions of measurements.

Gas sensors in an array can be contained in one compartment or in many measurement chambers. It depends on working mechanism of these devices. A sensor array with a set of separated chambers is preferred when individual sensors produce intermediate products during the sensing process which can affect responses of neighbouring sensors. In this case, the chambers have to be connected in parallel using a Teflon-tubing so that all sensors are simultaneously exposed to the gas mixture of the same composition. The compartment has to be airtight and made of chemically resistive material. The measurement chamber is provided with gas inlet and outlet. Geometrical parameters of this compartment such as volume, shape, and the inlet-outlet position influence the development of the fluid flow and the gas concentration distribution inside the chamber. Therefore, they must be considered. To avoid effects due to non-uniform gas concentration in the measurement chamber, it appears reasonable to place the sensor in the gas stream, very close to the point of injection. Similarly, the concentration transient may be reduced to a value substantially smaller than the response time upon reducing the chamber volume and increasing the volume flow rate through the chamber. That is based on hypothesis that the concentration transient is not much longer than the ratio between chamber volume and volume flow rate.

A sensor array may consist of many, closely placed sensors. In that configuration, neighbouring sensing elements can interact and interfere with each other. This phenomenon is often described as "cross-talk", and the response of the individual sensor depends not only on the gas under test, but also on its neighbours. From signal processing perspective, cross-talk can be considered either as random noise or as noise which has a certain pattern. The sources of the cross-talk can be different e.g. thermal, electrical, electromagnetic and architectural. This effect is considered as a local one and hence it may be removed by a suitable sensor array design.

The chemical and physical conditions directly around the sensor are also very influential. The sensor response may depend on the temperature of gas under test and the surrounding's. Thus, thermal conditions around the sensing elements should be controlled. This requirement can be realised e.g. by the installation of the sensors into a thermostatic chamber that keeps the temperature of gas at a constant level. Additionally, some of the sensors suffer from the relatively large power dissipation. Therefore, thermal isolation of sensor array is a key issue. In this way, the power consumption by the measurement device may be lowered.

The sensor array construction is strongly determined by the applied technology. Different sensor technologies are used in today's commercially available sensor systems. Recently, considerable interest has arisen in micro-electro-mechanical systems (MEMS). They are fabricated by the thin film and micromachining techniques (Faglia *et al.*, 1999; Vergara *et al.*, 2006). This technology meets the main requirements for gas sensor devices such as: small size, high sensitivity in detecting low concentrations, good selectivity, short response time, long term stability. The additional advantages such as low power consumption, possibility of on-line operation, temperature controllability, small size, low cost, easy realization of sensor array and possibility of on-chip integration with micro-electronics and low-cost fabrication make them attractive for consumer applications. Sensitive layers deposited on micro-machined substrates are the most promising structures.

3.2.3 Properties of individual sensors

The measurement abilities of the sensor array are strongly dependent on the measurement characteristics of the individual sensors, which are determined by their operating principles. The sensing elements in the array are not dedicated to any given analyte. They are chosen to respond to a number of different substances or classes of chemicals. With regard to the requirements imposed on the individual sensors that make up the arrays, the ideal sensing element integrated in the array should fulfil several criteria: good accuracy, resolution, precision and repeatability, high sensitivity towards target gas, medium selectivity (they must respond to different constituents of a tested sample), low sensitivity towards humidity and temperature, short response and recovery time, low detection limit, wide span and dynamic range, linearity, low drift and noise, robustness, durability, stability and sensitivity in hostile environment. Additionally, weight, dimension, power consumption, thermal capacity and thermal isolation, design and housing of the sensors are used to evaluate performance of these devices. It is difficult to achieve all of these requirements in practice.

Sensor arrays can include sensing elements based on different operating principles. Usually, they belong to one of the following classes of gas sensors:

- optical (absorbance, reflectance, fluorescence or opto-thermal effect are used for the measurement);
- electro-chemical (These devices are divided to potentiometric and voltammetric sensors. The last group includes amperometric devices);
- electrical (chemically sensitized resistors, diodes and field effect transistors);
- mass sensitive (piezoelectric sensors);
- magnetic for oxygen measurement;
- thermometric and calorimetric (thermal conductivity sensors, catalytic sensors).

Among different kinds of sensors that may be used in sensor arrays electro-chemical devices, mass sensitive sensors, metal-oxide-semiconductor field-effect transistors and chemi-resistors (semiconductor sensors) are especially interesting.

Electro-chemical sensors are the largest group of sensing devices. The important advantage of the first group of sensors is a linear range of the response. Usually, it is greater than 10^4. These devices are almost insensitive to changes in relative humidity (RH). They also respond to quite different chemical species than chemi-resistors or mass-sensitive sensors. The electro-chemical sensors respond only to gas that has electro-active properties. Therefore, they are not sensitive to some common classes of compounds, such as saturated hydrocarbons. These sensors will only increase the noise of the array during analysis of a sample containing these compounds. The sensing mechanism of these devices is controlled by the nature of the electrode and electrolyte as well as the thermodynamic potential of the sensing electrode, the rate of diffusion, solubility in the electrolyte, and the number of electrons produced per molecule. In general, the active Pt catalyst will record a signal for any electro-oxidizable gas or vapour (CO, SO_2, H_2S, NO_x, and EtOH). The NO_2 sensor is operated with an Au electro-catalyst at more cathodic potentials such that only electro-reduceable gases interact such as NO_2, Cl_2, and few others. Usually, electro-chemical sensors are optimized for a single analyte like: CO, H_2S, NO, NO_2, SO_2, Cl_2, O_3 or hydrazine. In this form, they are used in a sensor array. The measurement characteristics of electro-chemical sensors may provide complementary capabilities to sensor array.

Mass sensitive sensors (piezoelectric devices) transform the mass change at a specially modified surface into an alteration of a property (the resonant frequency) of the support material. The mass change is caused by an accumulation of the analyte on the adsorbent layer which is composed of stationary phases used in gas chromatography or by an accumulation in supramolecular host molecules. This group of sensors is split up into:

- quartz crystal microbalance (QCM) based on measuring the frequency change of the quartz oscillator plate caused by adsorption of a mass of the analyte at the oscillator;
- surface acoustic wave (SAW) devices depend on the modification of the propagation velocity of a generated acoustical wave affected by the deposition of a definite mass of the analyte.

Mass sensitive sensors present very good low detection limit. However, their selectivity is low. The sensitivity of the QCM and SAW devices to a particular compound depends on the type of sensitive membrane. These sensors can detect a broad spectrum of chemical species due to the wide range of gas sensitive coatings available. However, these devices suffer from poor SNR. It is caused by the high frequencies, at which SAW sensors are operated or surface interferences in the case of QCM sensors. The circuitry required to operate mass sensitive devices is complex and expensive.

The metal-oxide-semiconductor field-effect transistor (MOSFET) is a transducer device, which transform a physical/chemical change into an electrical signal. The sensing mechanism of MOSFET sensors is based on the principle that the threshold voltage of this device changes upon interaction of the gate material, usually catalytic metal, with certain gases due to the corresponding changes in the work functions of the metal and the oxide layers. These changes are caused by the polarization of the surface and interface of the catalytically active surface. Gas sensing MOSFETs are produced by standard micro-fabrication technique or in the micro-machining technology. In the last case, a hybrid suspended gate FET (HSGFET) sensors are fabricated. Gas sensing MOSFETs have a number of advantages, which result from the CMOS technology. The reproducibility of these devices is quite good. They are small and cheap. Gas sensing MOSFETs also have disadvantages. For example, they suffer from baseline drift and instabilities depending on the sensing material used. Additionally, the properties of these devices are dependent on the gas flow and operating temperature. Therefore, control of the surrounding environment is required.

Chemi-resistors are very often described as the semiconductor sensors, due to the working principle of these devices, which is based on the variation of the sensing material conductivity in presence of oxidizing and reducing gases. Different organic materials e.g. conducting polymers or metal oxides e.g. ZnO, TiO_2, WO_3 and SnO_2 are used in this sensor technology. Conducting polymers offer many advantages over other sensing materials. An array equipped with polymeric sensors can obtain high discrimination ability because different polymers generate various levels of response to a given gaseous sample. A wide range of polymeric materials is available on the market. They are relatively inexpensive and easy to prepare. Sensors fabricated from polymeric materials can be used in conditions of high relative humidity. They show highly linear responses to many gases. These devices are operated at room temperature. Therefore, power consumption is very low. It means that polymeric sensors can be applied in portable, battery powered equipment. Signal conditioning circuitry required for these sensors is relatively simple as only a change in resistance is measured. The main disadvantage of this type of sensors is aging, which causes drift.

Semiconductor metal oxide gas sensors are considered as one of the basic technologies for the array applications. It results from the advantages of these devices e.g. high sensitivity, short response and recovery time, durability, small weight and dimensions, a real simplicity in function, large number of detectable gases (possible application fields are wide), low cost, flexibility in production. Although semiconductor gas sensors are widely used as invaluable safety devices for the detection of methane, propane or carbon monoxide gas, it is known that these devices suffer from a number of shortcomings such as: lack of selectivity, nonlinearities of sensor's response, long-term drift, variations in the initial resistance, sensing material poisoning and aging, relatively high operating temperature (> 300°C), hysteresis, sensitivity towards humidity and temperature. Therefore, the important goal in the development of chemiresistors is to improve the measurement characteristics.

This requirement can be achieved in different ways since semiconductor sensor response to target gas depends on many factors. For example, the improvement of sensitivity, selectivity and response time of theses devices can be reached by:

- the selection of a semiconductor with the suitable intrinsic properties (metal oxides e.g. SnO_2, ZnO, TiO_2 or organic materials e.g. phthalocyanines, conducting polymers);
- the addition of catalysts and promoters or more specific surface additives;
- bulk/surface dopings;
- the choice of a suitable physical parameters of sensing layer (thickness of the gas-sensing entity, internal porosity, layer microstructure, crystallinity and crystalline structure, grain size, grain-grain contacts, surface morphology);
- the application of different technologies of the sensing material deposition and formation on the substrate;
- the aging and the preconditioning procedures;
- the modification of sensing layer, especially its surface; application surface coatings e.g. membranes and a diffusion filter layer such as SiO_2 on top of the semiconductor;
- the selection of a suitable material and geometry of the substrate;
- the appropriate choice of material and setup of the electrodes;
- the choice of the sensor structure and enclosure.

3.3 Working parameters

The measurement characteristics of sensor array can be controlled to some extent by working parameters. Mechanism of gas sensing causes that operating conditions of chemi-resistors are primarily determined by the temperature, flow rate of sample, gases partial pressures or potential (bias). These parameters may be constant or time-varying in course of the measurement. The alteration may be predetermined (programmed), controlled by the sample and the sensor or accidental.

3.4 Operation mode

In principle, the term operation mode means a manner or a way employed to operate a device. In practice, it is a description of conditions under which analytical equipment works. Usually, the operation mode is characterized by an applied procedure, sensor environment, method of sensor response (output signal) measurement and working parameters.

Operation mode may affect the performance characteristics of semiconductor sensors since it determines the state of these devices during measurements. Generally speaking, a gas sensor can be in a steady or unsteady state. The first case is obtained in stable conditions that do not change over time. The sensing material is in equilibrium with a surrounding gas. In commercial instruments, the sensor responses in this state are preferred up to now, because the output signals are easily measurable and time-independent. The steady-state of sensor's signal as input for discriminators are easier and less complicated for processing and data analysis. Additionally, problems with unstable flow, pressure and temperature of gas sample, which are encountered at the beginning of exposure are minimised. However, when a sensor is in the steady state a lot of information about tested gases is lost. First of all, there is inaccessible information originating from the kinetics of processes that cause sensor response. In many applications of chemical sensors, information can be gained not only from a steady-state value of the sensor signal, but also from the kinetics of the response. Hence, an alternative approach has received much attention in recent years. It is based on sensor responses, which are recorded when the device is in an unsteady state. In this case, the sensor is not in the equilibrium with a chemical environment. The sensor response in the unsteady state is a time-dependent (temporal or transient). It conveys information, which is particularly useful to an enhancement in the discriminating ability of chemiresistors. For that reason, the semiconductor sensors in this state have received much attention in recent years.

The operating mode is often classified as a static or dynamic. Static mode means that any quantity acting on the sensor is constant with time. In other words, the device is exposed to the gas under test in stationary conditions. The dynamic mode is based on alteration of working parameters or operating conditions with time. A special variant of this operation technique is based on signal modulation. In this mode, the alteration of the quantity acting on the sensor or sample is characterized by frequency and amplitude. The dynamic mode operation techniques can be grouped into four categories:

- modulation of measurement parameters such as potential (bias) or frequency (impedance spectroscopy) - AC operation mode (Amrani *et al.*, 1992; Gutierrez *et al.*, 1991, 1992);
- modulation of the working temperature of the sensors (Gouws & Gouws, 2003; Gutierrez-Osuna *et al.*, 2003; Llobet *et al.*, 1997; Vergara *et al.*, 2007);
- modulation of the gas flow (Barbri *et al.*, 2008);
- modulation of the gas concentration (Gouws & Gouws, 2003; Llobet *et al.*, 1997).

In the AC operation mode, a periodic waveform (e.g. a sinusoidal) is applied to the sensor input as a reference voltage instead of a fixed DC power supply. The voltage in the sensor heater is kept constant. Gas discrimination is enhanced in AC operation mode, by taking measurements of different electrical parameters (such as sensor capacitance, conductance or dissipation factor) at different frequency values of the reference voltage generator. This offers several advantages. For example, SNR is usually better, because narrow band amplification (either by filters or lock-in techniques) can be implemented and the 1/f noise component is less significant. Thus lower detection limits can be achieved.

In the past few years, modulating the working temperature of metal oxide sensors has been one of the most used methods to enhance sensor selectivity (Gutierrez-Osuna *et al.*, 2003).

This mode consists in altering the kinetics of the sensor response through changes in the operational temperature of the device. It is well known that when the working temperature of a gas sensor is modulated, the kinetics of the gas-sensor interaction is altered and this leads to characteristic response patterns. Temperature modulation can be grouped into two broad categories:

- thermal transients;
- temperature cycling (oscillation).

In the thermal transient approach, the heater voltage of the sensors heater is driven by a discontinuous step function (e.g. a heater supply is switched from a high voltage to a low voltage) or a pulse signal that is used to drive the heater voltage. The discrimination is performed in the transient response induced by the fast change in temperature. In the temperature cycling, the sensor heater is driven by a continuous function. Different modulating waveforms (e.g. ramp, sinusoidal, triangular, saw-tooth and asymmetrical square waves) can be used for temperature oscillation. The semiconductor ceramic sensors deposited on a substrate with high thermal inertia (e.g. TGS sensors) do not seem the best candidate sensors for an effective temperature cycling. Under these conditions extremely low modulating frequencies are used, which results in long, impractical, measurement times. With the development of the micro-system technology, the availability of micro-machined substrates for metal oxide gas sensors implied that sensors could have their operating temperature modulated in a more efficient way.

The effect of flow modulation can be analysed from two points of view. Firstly, the concentration of analytes at the surface of sensors is modulated. Secondly, fast periodical flow changes result in periodical cooling and heating of sensors' surface. Therefore, specific response patterns, which are characteristic of the analytes present, develop. The method can be easily adapted to both static and dynamic headspace sampling strategies. A wide range of concentrations and contaminants have been tested which confirms that flow modulation allows for a reliable identification of different vapour species.

Concentration modulation produces an output signal that contains information on the dynamic adsorption and desorption processes. The transient signals in this mode are generated while the controlled modulation of a sensor input parameter is performed. The concentration modulation may be achieved by means of fast-switching valves. The frequency spectrum of these transient signals should be a source of information containing details on the dynamics of the interaction process and have the potential for analyte identification.

Usually, transient characteristics are obtained by operating sensors at different temperatures. In particular, modulation of this working parameter is used to reduce long time drift effects and diminish cross-sensitivities. Unfortunately, this method requires a temperature control and additional equipment. The operation based on altering the target gas partial pressure to get unsteady state of the sensor is relatively rarely tested and applied in practice.

3.4.1 Stop-flow operation mode

In the last years, considerable efforts have been made to use sensor dynamics as a source of multivariate information aimed at an enhancement in the discriminating ability of poorly-selective metal oxide gas sensor arrays. The detection of reducing gaseous substances, e.g. volatile organic compounds by chemiresistors like Taguchi gas sensors (TGS) is caused by a change in conductivity of semiconductor. This effect is induced by the interactions between gaseous molecules and sensing material. The mechanism of chemiresistor operation is complex. It is controlled by several surface phenomena such as: diffusion of target gases into the sensing layer through pores, adsorption/desorption, catalytic reactions and followed by a charge transfer between the adsorbate and surface of sensing material leading to a measurable change in electrical resistance of semiconductor. These phenomena are directly or indirectly affected by factors, which function as working parameters of sensors.

Usually, this group of parameters includes temperature, potential (bias), sample flow rate and partial pressure or concentration of target gas. The output signal of gas sensor depends not only on an absolute value of working parameters, but also on measurement conditions. It results from a nature of the mentioned phenomena. Measurements in stationary conditions reflect mainly the thermodynamics of interactions between target gas and sensing material, whilst results obtained in dynamic conditions contain information about kinetics of adsorption/desorption, chemical reactions and charge transfer. Each of these processes has its own characteristic time constant, which is dependent on properties of sensing material and gas molecules, especially their molecular weight and shape, electron affinity, ionization potential, diffusion coefficient. Hence, the informative contents of sensor response measured in dynamic and static conditions, in steady or unsteady state are quite different. In our research work, it is assumed that the diversification of working parameters and operating conditions may cause increase in sensor ability to distinguish particular substances. Hence, an operation mode is proposed, which offers a significant alteration of the listed factors during measurements. The mode is called stop-flow. This approach consists of a few predetermined and controlled steps. The main factor, which defines each of these stages, is the alteration of exposure conditions, e.g. the sample flow rate. Other working parameters may also be changed during particular steps. During measurements sensors are either in steady or in unsteady state, and both dynamic or static conditions of operation are encountered. The stop-flow operation mode was characterized with regard to its application for measuring system based on the sensor array (Maciejewska *et. al*, 2009, 2010; Szczurek *et al.*, 2010).

The responses of sensors in the stop-flow operation mode are obtained using the procedure consisting of three successive stages. The first step is the dynamic exposition. The gas sample is allowed to continuously flow through sensor chambers. The gas flow rate through the experimental setup is set to a defined value and it is kept constant. The dynamic exposition takes place. After this step, the gas flow is stopped and static exposition to the test gas is performed. The third stage could be described as a recovery process. The gas line and chambers are cleaned with a stream of pure air. The air flow rate through the system is again constant. The duration of each step is defined. They usually last for several minutes each and these times add up to the length of entire measurement. The regeneration of sensors is continued afterwards, until readouts from these devices reached the level as before the dynamic exposition. In our studies, a reversible change in the resistance of sensors was observed in all cases.

The typical output signals of TGS sensors obtained in stop-flow operation mode are presented in Fig. 1.

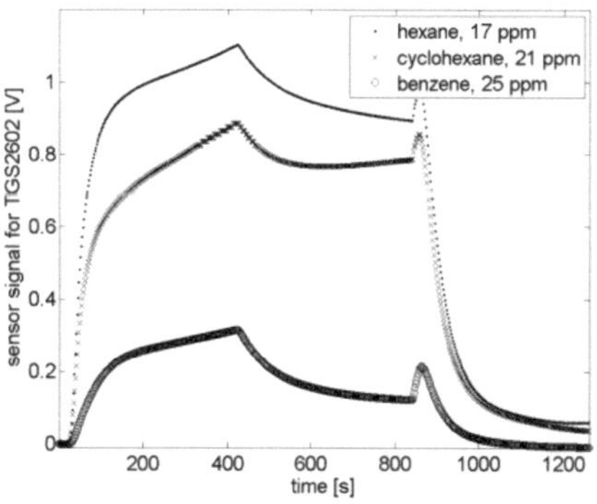

(a) Three substances measured with one sensor

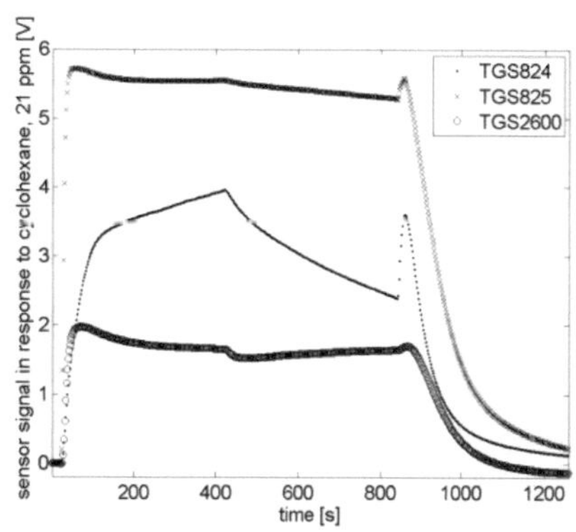

(b) One substance measured with three sensors

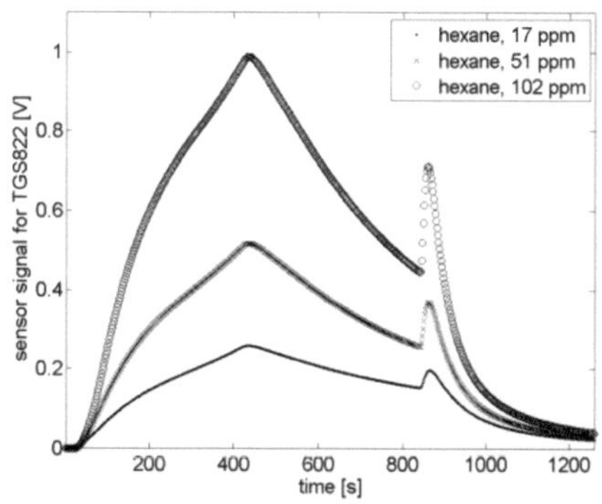

(c) One substance at various concentrations measured with one sensor

Fig. 1. Typical output signals of TGS sensors obtained in stop-flow operation mode.

Each of the output signals shown in Fig. 1 can be split into three main parts. They correspond to different conditions of operation during particular stages of the stop-flow operation mode. One can characterize the signal in respect of this partition.

As a consequence of the applied measurement procedure, after sample injection into the apparatus, some time is needed to exchange the atmosphere in the chambers. During this time interval the gas concentration is continuously changed, until it reaches the maximum value. As a result, the sensor signal changes, first rapidly than at lower rate until quasi-constant value is reached. It is observed that single minutes were sufficient to attain the gas equilibrium in the atmosphere surrounding sensors. The transient output signal upon exposure of sensor to a stream of gas sample was caused by the change in analyte concentration and the diffusion-limited processes which take place in the porous structure of the sensing material. It is known that the conductivity changes in semiconductor sensors depend on: the transport of the reactive species into the sensor, the diffusion of the gas molecules inside pores of the sensing material, adsorption and desorption, the catalysed redox reactions on the surface of the sensing layer (mainly their kinetics) and the electrical/electronic effects in the semiconductor. The sensor response resulting from those processes is strongly related to the concentration and chemical properties of VOCs. For that reason, the transient output contains information about the tested gases.

The second part of sensor response starts when the gas flow is stopped. At this moment, the static exposure begins. The sensor response depends mainly on desorption and diffusion processes. Also the change in sensor temperature has its influence on it. The oxidation of organic compounds at the semiconductor surface and the lack of gas sample flow inside chambers cause the concentration of analyte to diminish. It results in the decreasing output signal. It is shown in Fig. 1 that the second phase, just like the first one is featured by versatile sensor response patterns.

In the third phase, upon the recovery process due to the removal of organic compounds from sensor chambers, the drop of measured output signal is observed. The rate of decrease is sensor-dependent. It is also influenced by the analyte and its concentration. The initial surge of signal in this phase may be observed with its magnitude and duration depending on the dead volume of the gas line.

The examination of sensor signal shown in Fig. 1 reveals that coupling flow and no-flow conditions in one operation mode of sensor system shaped the response pattern of sensor array. Each, dynamic and static exposure yielded transient responses with characteristic shapes and time constants which are affected by the composition of the atmosphere surrounding sensors. Therefore, the chemical information contained in measurement could be useful for distinguishing different gases.

3.5 Signal processing

The response of the sensor array is measured and converted into an electrical signal (e.g., a voltage). This function is realised by means of interface circuits. In sensor techniques, the term "signal" can be understood as a sequence of measurements that are related by some ordinal variable such as time. In the beginning, the raw measurements are converted from analogue readings to a digital signal that can be interpreted by a computer. The digitised signal is processed online or stored for future analysis. The signal pre-processing stage operates on the gas sensor outputs in a way that improves the overall pattern analysis performance. It can be achieved by extracting parameters that are descriptive of the sensor array responses. Thus, the raw data are transformed into a characteristic feature vector. Although signal pre-processing will depend on the application, a series of steps are commonly carried out. Signal pre-processing serves various purposes and consists of the following operations:

- baseline manipulation;
- compression;
- normalization.

The final effect all of these operations is a set of features arranged in a form of an initial feature vector.

Baseline manipulation procedures transform the sensor response relative to its baseline to minimise the effects of temperature, humidity and short term drift. Three baseline methods are commonly used: difference, relative and fractional. The selection of a proper baseline manipulation technique is highly dependent on the sensor technology and the particular application.

Sensor array signals are represented by a set of data. While taking a measurement with the multi-element sensor array operating in dynamic conditions a lot of data is acquired. The number of collected data depends on the size of sensor array, the duration of time response and sampling rate. Usually, not all data are necessary to form a distinct pattern of measured gas. In practice, some of them contain no valuable information or are strongly correlated with other variables. Additionally, an analysis of a large set of data may cause serious calculation problems. The elimination of data, which contain no valuable information in the specific application or which show large correlation to another data within data set, may be advantageous. Thus, it is necessary to reduce the number of variables, which are analysed, resulting in reduced acquisition time, improved selectivity and recognition accuracy. This is aimed at using only relevant data in feature vectors for pattern recognition purpose.

Various compression algorithms are employed in the sensor technique. The standard procedure is to select the steady-state response of a sensor. To extract additional information from the transient response, a few methods have been proposed e.g.:

- model fitting;
- ad-hoc parameter extraction;
- sub-sampling.

A signal can be defined as an information-conveying function of one or more independent variables. Sometimes, this function can be described in an analytical form. For example, the transient response may be modelled as the sum of exponential functions, also using polynomials, autoregressive and state space models and their estimated parameters form the initial feature vector. Exponential curve-fitting methods can result in nearly lossless compression of the sensor transients but are computationally intensive. For these reasons ad-hoc parameter extraction and sub-sampling methods are more commonly employed.

The ad-hoc parameter extraction approach relies on a compression of sensor array response down to a few single parameters without fitting sensor signal with any kind of function. They are usually easy to calculate. Each of them should contain information about gases which are in the multi-component mixture. Different descriptors may be extracted from the transient response curves e.g. response value of initial saturation or in steady-state, final or maximum response (deep saturation), average of several points or of the whole signal, the response rate of sensor to tested gas (transient slope), pulse heights, derivatives, integrals and time constants (e.g. time-to-threshold) They are used as coordinates of feature vectors or elements of fingerprints for pattern recognition purpose. Unfortunately, in the model fitting and ad-hoc parameter extraction approach much information about chemicals may be lost or inaccessible. This refers in particular to the information, which is locally present in the signal and therefore any representation, which refers to longer parts of signal shadows it. Conceptually, it fits considering sensor signals as a set of discrete and separate information sources. Solutions employing model fitting or parameter extraction technique may require complicated analysis of the whole dataset. Thus sub-sampling has received much attention recently. In this method parameters describing signal are exchanged by a set of signal sub-samples. This term means samples drawn from a large sample. The set of sub-samples is a compact representation of the all sampled data. The complicated form of the sensor signal causes that sub-sampling is an attractive method of data compression in the stop-flow operation mode.

3.6 Data analysis

In principle, a sensor array with broad applicability has to face numerous tasks of gas assessment. In general, they are of qualitative or quantitative kind. At the level of sensor array data analysis these tasks are addressed as pattern recognition problems. Classification and regression tasks are considered respectively. Data analysis involves defining pattern recognition problems, which correspond to gas assessment problems, and finding best ways to solve them. The main elements in this process are: patterns, feature space and pattern recognition.

3.6.1 Patterns

The feature vector is the basis for building a pattern which represents a measured gas. Pattern is obtained by assigning values to features, which are elements of feature vector. As features are parameters of sensors signals, their values are taken from sensor array measurement data. For developing a targeted sensor array, calibration patterns are used. Calibration patterns come from measurement data, which are collected while exposing sensor array to known gases. Therefore, it is straightforward to assign a calibration pattern with a discrete label, which indicates its membership in a particular class, respectively to the category of the measured gas. The calibration pattern may also be assigned with a value of continuous variable, which refers to a quantitative parameter of calibration gas, e.g. its concentration. The set of calibration patterns which refer to various gases and the associated label vector/concentration vector is the multivariate calibration data. Using this data, supervised classifiers are trained to define decision boundaries between patterns, which belong to different classes. Also regression methods are employed for setting the transformation between the matrix of patterns and the corresponding concentration vector. This process is called a supervised learning. The obtained pattern recognition models are further used for prediction purposes, regarding patterns associated with unknown gases. In our research on sensor arrays, it is favoured to work with many patterns representing the same test gas and many pattern recognition models respectively.

3.6.2 Feature space

Pattern recognition proceeds in feature space, which is defined by the selection of sensor signal parameters to become features. Many feature spaces are usually available for solving pattern recognition tasks based on sensor array data. The accessible spectrum of features depends on the pool of sensor signal parameters to select from. Particularly extensive sources of features are signals obtained when using techniques to increase their information content e.g. the stop-flow operation mode. The effectiveness of solving pattern recognition task is related to the feature space, where it is performed. Actually, it is determined by the mutual distance of sensor array response patterns, which represent various gases in that space. Pattern coordinates are set by features. Therefore, the selection of features decides about the pattern layout, and makes it different in different feature spaces. It is imaginable that there exist a feature space(s), which offers the best solution of gas assessment task in terms of the corresponding pattern recognition task, as compared to other spaces, due to the most favourable layout of patterns in this space.

In our work the idea of solving each gas assessment task is to use the feature space(s) where the associated pattern recognition problem is solved best (Szczurek *et al.*, 2011). For realisation of this approach it is good to propose a way for defining many significantly different feature spaces. There are a number of factors, associated with construction and operation of sensor arrays, which cause the differences among features. Looking at the measurement data from a sensor array, they are most readily represented by the sensor dimension and the time dimension accordingly (see Fig. 2). The variability associated with the sensor dimension is commonly exploited for pattern recognition in sensor arrays. It actually is a part of the sensor array concept. The realisation is performed using the parameters of signals of many sensors as features. Recently, attention is paid to the utilisation of time dimension. In general, there are two ways to include factors revealed in time domain for shaping feature space. The direct way consists in using parameters, which synthetically represent entire sensor signal or its substantial part, as features. The best candidates are parameters of mathematical models used for fitting sensor signals. The other approach uses results of single measurements, which the sensor signal is composed of, as features. Here the time domain enters the feature space indirectly, as a set of variables associated with different time moments of the sensor signal. As mentioned before, the second strategy is called sub-sampling. In principle, the concept of sub-sampling refers to a single sensor signal. We proposed its extension to the collection of signals from the sensor array. In this way a uniform framework is proposed, which allows to build versatile feature spaces. They may account for all different sorts of factors, which are associated with sensor array construction as well as operation and which cause differences among features. Considering the kind of feature used, it actually is a data driven approach.

In our strategy, the output signal of sensor array operating in the dynamic conditions, was a result of a sequence of discrete measurements. The data obtained in course of exposure may be arranged in the following matrix form:

$$\mathbf{R} = \begin{bmatrix} r_{1,1} & \dots & r_{1,j} & \dots & r_{1,n} \\ \dots & \dots & \dots & \dots & \dots \\ r_{k,1} & & r_{k,j} & & r_{k,n} \\ \dots & \dots & \dots & \dots & \dots \\ r_{m,1} & & r_{m,j} & & r_{m,n} \end{bmatrix} \quad \text{time dimension}$$

sensor dimension

Fig. 2. Sensor array data arrangement. Sensor and time dimension are associated with main groups of factors, which cause variability in feature space.

The matrix dimension is mxn where, m is the number of measurements during single exposure to gas mixture, and n is the number of sensors in the array. In this notation k, k=1,…, m corresponds to different time points of the exposure time interval, while j, j=1,…,n indicates sensors in the array. A single matrix element is $r_{kj} = r_{kj}^{gas} - r_{0,j}$ where r_{kj}^{gas} is the output signal of jth sensor to gas mixture, measured in k^{th} time point of exposure, and $r_{0,j}$ is the baseline signal. This data arrangement may be fixed. In that form, it corresponds to a strictly defined sensor array setup and measurement procedure and may be considered as a template of the response of sensor array to a gas mixture.

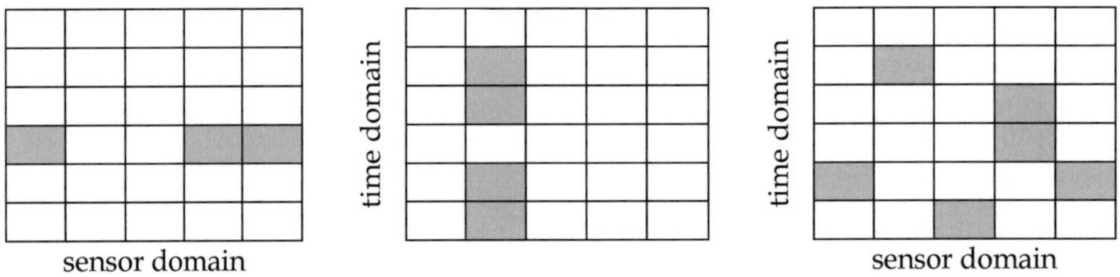

Fig. 3. Location of feature vector elements in sensor array data matrix. The following sources of features variability are represented in the feature vector: (a) sensor kind, (b) factors acting in time domain, (c) both, sensor kind and factors acting in time domain.

In our work the element r_{kj} of data matrix **R** is considered a feature. Within the framework of extended sub-sampling, in the feature vector there may be represented:

1. variability associated with the senor kind,
2. variability exhibited in time domain,
3. both kinds of variability jointly. These three cases are illustrated in Fig. 3.

The approach allows for generating different feature spaces. Additionally, the number of feature spaces is substantial. Considering an array composed of *n* sensors and sensor signal composed of *m* discrete measurement results, the number of feature spaces is:

$$N_{fs} = \sum_{k=1}^{n \cdot m} \frac{(n \cdot m)!}{k!(n \cdot m - k)!} \qquad (1)$$

where *i* is the number of variables in the feature vector.

In our recent works, the feasibility of the approach regarding feature vectors is shown where the differences among features are caused by the sensor kind. It is indicated that the contribution of this factor for pattern recognition is variable upon varied conditions of sensor array operation and the resulting sensor states (Maciejewska *et al.*, 2009; Szczurek *et al.* 2010, 2011). The steady state, which is commonly favoured in commercial devices based on a sensor array is shown to provide equally or less valuable feature spaces, as compared to other sensor states. This gives a reason for further investigation of factors, which act upon a sensor array in time domain. Moreover the interaction of these factors and sensor variability builds another perspective. These can all be exploited for extending the applicability of a sensor array.

3.6.3 Selection of best feature space

With the emphasis on solving gas assessment task using the feature space(s) where the corresponding pattern recognition problem is solved best, attention is paid to the search for an appropriate feature space. In principle, it can proceed in two ways, via feature extraction or feature selection. Feature extraction is achieved by transforming entire initial feature space into a new one. The main advantage of this transformation consists in compressing valuable information into new features, which are fewer than the original ones. Often, new features also have other advantageous properties. For example, when using principal component analysis (PCA) as a method of extraction, new features are orthogonal. Furthermore, features resulting from independent component analysis (ICA) are statistically independent. The independence or at least orthogonality is beneficial when applying regression analysis. However, the extracted features lack the direct physical interpretation, which is available for the original ones. Also, the number of transformations of the original feature space is limited.

In our work the search for an appropriate feature space is done by means of feature selection. The goal is to find a set of original features, which maximises the information content or the predictive accuracy criterion. This is synonymous with finding best space(s) of original features. The selection involves two aspects:

1. search through feature sets,
2. assessment of feature sets. In classical perspective the sets are actually subsets of one, original feature set.

Although substantial workload and time may be required at this stage, in this way we avoid one additional transformation of feature space before proceeding to pattern recognition stage. Additionally, much bigger flexibility when dealing with feature vector composition is retained.

Considering search strategies, the one which guarantees arriving at globally optimal solution is the exhaustive search. However, the number of possible subsets grows exponentially with the number of features. The approach is impractical for spaces composed of more than 20 features. To obtain a solution in a reasonable time, the search strategies are used which examine the feature space in a more efficient way. They may be grouped into: heuristic, deterministic and stochastic. The heuristic approach is attractive as a means of initial reduction of feature space, and ultimately shortening the search time. It utilises the expertise of the researcher, especially regarding information value of single features. Deterministic search strategies are mostly algorithms of local optimization, except for the branch and bound technique. Therefore, they are claimed suboptimal. However, many of them are attractive due to fast convergence. This group is dominated by variants of sequential search. Search strategies, which are considered nearly optimal, belong to the last category. Among the stochastic methods, genetic algorithms and simulated annealing are the most acknowledged ones. In these algorithms the random element is involved in the selection of feature sets. For that reason they are less prone to stucking in a local minimum. However, the algorithms are also computationally more demanding and time consuming than deterministic approaches.

The assessment of feature space is usually carried out using one of three following approaches: filter, wrapper or embedded. The first approach is free from the context of pattern recognition method. It utilises the assessment criteria like e.g.: the distance between object categories, interclass vs. intraclass pattern variability, information content measures. Due to using so general criteria it has high cognitive value, in particular as a means of assessing individual features. Filtering is willingly employed for sensor array optimisation on a feature by feature, univariate basis. Regarding multivariate feature assessment and pattern recognition context, filters have smaller practical value than other approaches. The wrapper approach allows for the assessment of joint performance of pattern recognition method and feature space in the context of particular pattern recognition task. The result is less general. However, the method is well suited for finding best feature spaces for solving gas assessment task represented as a pattern recognition one. The approach favours simple classifiers like linear discriminant analysis (LDA), k-nearest neighbours method (k-NN), because the time required for model learning is the limiting factor just like the number of feature spaces to search through. In case of using more time consuming methods e.g. ANN, SVM, feature spaces pre-selection is usually done in a simpler manner. In recent years the embedded approach is slowly gaining interest in the sensor array community. The name comes from embedding the strategy of feature selection in the process of classifier development. The most serious candidates in this group of methods are random forests.

Considering the concept of finding best feature space for each kind of gas assessment task, which is realised with a sensor array, the wrapper and embedded approaches are favoured for feature space evaluation. So far in our works the first one was examined in combination with the heuristic pre-selection. Using heuristics, the feature spaces were preselected in which the differences among features came from the differences among the sensors at the particular moment of their exposure to the test gas. The rule was applied to the data provided by the sensor array composed of 15 sensors, which was operated in the stop-flow operation mode. As a result, over a thousand of 15-dimensional feature spaces were obtained. It was demonstrated that the complete search of those feature space could be performed in an acceptable time. To a large extent it was achieved due to employing fast learning pattern recognition methods. Regarding feature set assessment criteria, the classification success rate was used in the case of qualitative problems and the error of concentration prediction (RMSE and/or MRE) was applied for quantitative problems. The feature space was considered best for solving a particular qualitative problem of gas assessment if the corresponding pattern classification task was solved with null classification error rate. The feature space was considered best for solving a particular quantitative problem of gas assessment if the projection of feature set into corresponding quantitative variable resulted in a concentration estimation error, which was lower than the threshold of several percent. In our research a number of gas mixtures were considered to investigate the assessment problems (Maciejewska *et al.*, 2010; Szczurek *et al.*, 2011). The results show that many best feature spaces are available for each of them. Usually, there are more than several hundred ones in single case. It is meaningful that numerous low dimensional spaces (3 and even 2 dimensional) were found to fulfil the criteria of best feature space.

3.6.4 Pattern recognition

In the research work on sensor arrays we retract from looking for the pattern recognition method, which performs best in an imposed feature space in favour of searching for the feature space(s) where the particular pattern recognition method performs best. The first approach is the conventional one. In general, it allows for highly multi-dimensional feature spaces and it favours nonlinear pattern recognition methods. The most popular ones are artificial neural networks (ANN) and support vector machines (SVM). ANN are available in many kinds: multi-layer perceptrons (MLP), radial basis function neural networks (RBFNN), probabilistic neural networks (PNN), Kohonen maps, just to mention the most frequently used ones. They are suitable for solving linear as well as nonlinear problems. Additionally, classification as well as regression tasks may be approached with ANN. One of their drawbacks results from the fact that models have many parameters to be adjusted and this requires substantial calibration data sets. SVMs in turn offer very good classification results in the cases when data sets are small and the classification problems are highly nonlinear.

They are competitive as compared to ANN regarding the classification success rate, the number of user defined parameters and the time required for learning. However, they are not used for mapping features into continuous variables.

It is most frequently assumed that sensor array pattern recognition problems are nonlinear and therefore the above listed methods are preferred. However, there are many examples of successful pattern recognition using simpler, in particular linear methods, even if the feature space is not selected with using them in mind (Arnold *et al.*, 2002; Pardo *et al.*, 2006). In our approach simple classifiers and simple regression models are used. It is assumed that the probability of finding a feature space where patterns are linearly or at least nearly linearly separable is nonzero. Similarly, an assumption is made concerning the existence of feature set, which can be linearly transformable into the gas concentration. So far, linear and fast to train pattern recognition methods have been used in our research on senor arrays with broad applicability. These are: DFA for classification tasks and multiple liner regressions (MLR) for quantitative problems. They performed very well as feature space assessment methods in the framework of wrapper approach (Maciejewska *et al.*, 2010; Szczurek *et al.*, 2011). The exhaustive search of 15 dimensional feature spaces is relatively fast and it allows for arriving at solutions in an acceptable time. It was shown by the results of our research that there were available numerous feature spaces where qualitative gas assessment could be solved as linear pattern recognition task. The same was found regarding quantitative gas assessment. In both cases simple pattern recognition methods were sufficient.

3.6.5 Data analysis arrangement for test gas assessment

While designing sensor array with broad applicability it is particularly important to propose reliable and functional solutions regarding measurement data analysis. Our proposal is grounded on the ability to achieve best results of pattern recognition in low dimensional spaces, by means of simple classifiers and regression models. The key elements of this concept are:

- A sensor array is prepared for solving defined problems of qualitative and quantitative gas assessment. However, the list of tasks is not a closed one. It can be extended any time by the producer of sensor array. The aim of this possibility is to increase the universality of measurements.
- The problem of qualitative gas assessment is considered as the corresponding classification problem (prediction) and the problem of quantitative gas assessment is approached as the corresponding regression problem (prediction). They are solved in an earlier defined feature spaces. These are best feature spaces in a sense, which was discussed earlier.
- There is assigned one, parameterised classifier for solving a particular classification problem in one, feature space and there is assigned one parameterised regression model for solving a particular regression problem in one feature space.
- A structure called data record is used for keeping the information about one feature space and the associated classifier/regression model, both assigned for a particular qualitative/quantitative problems. A data record is composed of two parts. The first part defines the feature space by means of feature vector. In our approach features are elements of matrix R (see Fig. 2). The coordinates of feature vector are indicated using

the position (k, j) of elements in data matrix R. The second part of data record contains parameters of the associated classifier/regression model in a predefined order. The parameters of pattern recognition models are obtained in course of supervised learning, using calibration patterns.

- A number of data records are available for each qualitative gas assessment problem and it is so for each quantitative gas assessment problem. We propose that the set of data records addressed to a particular gas assessment problem contains between several and several dozen data records.
- Data records are defined by the producer of a sensor array and they are held in the library of the device. In our strategy, this operation is realised at a level of sensor array design, on the basis of calibration measurements. Responses of different sensors to standard gaseous mixtures (in accordance with a procedure applicable in the stop-flow mode) are considered. Data records may be updated in course of periodic calibrations.
- The qualitative gas assessment is provided as the result of majority vote of the committee of classifiers. Member classifiers operate on different patterns, which represent the same gas in different feature spaces, following the reference provided by the respective data records.
- The quantitative gas assessment is given as the median of predictions provided by the set of regression models. Models operate on different patterns, which refer to the same gas in different feature spaces, following the reference provided by the respective data records. Quantitative gas assessment, is preceded by the qualitative assessment unless the identity of gas is known.

Due to low dimensionality of best feature spaces and the simplicity (small number of parameters) of pattern recognition models, the memory requirements are very modest in the proposed arrangement. Also, the computational requirements are low, which offers that gas assessment is performed quickly, once the measurement is complete. The user interaction with the data analysis is very limited and restricted to the selection of gas assessment problem, which needs to be solved, from the library of the device.

4. Conclusions

Gas sensors already have a considerable share in detecting and monitoring the presence of hazardous gases in various atmospheres. However, for achieving their broad application, the ability to perform qualitative and quantitative assessment of chemical substances is required. Despite numerous positives like fast response time, small size and low price, gas sensors have some limitations, in particular regarding sensitivity and selectivity. The recognised way to overcome them and to achieve the requested functionalities is to use a sensor array. It is based on the idea that selectivity may be achieved by using a set of partially, but differently selective sensors. It was actually shown that the responses of such a set provides data patterns, which represent the measured gas in a nearly unique manner. The gas assessment is achieved by analysing the pattern with pattern recognition methods.

Sensor array design involves a number of aspects. The major ones are:

1. method of sampling and sample pre-treatment,
2. measurement characteristics of sensor array,

3. working parameters,
4. operation mode,
5. signal processing,
6. data analysis.

The improvement on any of these elements contributes to the overall performance of sensor device. Their careful consideration is especially important regarding design of sensor array with broad applicability.

Sampling is necessary for presenting samples to sensors. Two main kinds are active and passive sampling. While diffusion sampling is practically effortless, dynamic sampling requires more complex measuring setup. At that cost, the gas delivery is under control, lower gas concentrations can be measured and instrument calibration is easier. Therefore, the dynamic sampling is our choice for sensor array addressed to wide range of applications.

Performance of a sensor array is directly influenced by the selection of sensors and the construction of sensor array. Selection of sensors refers to the kind of sensors and the number of sensors in the array. Many kinds of sensors are applied in sensor arrays. The most popular ones are: electro-chemical sensors, mass sensitive devices, MOSFETs and chemi-resistors. Due to utilizing different sensing mechanisms, heterogeneous sensor arrays are more versatile. More popular, homogeneous arrays are easier to construct and operate. The measurement characteristics of sensor array are not proportional to the number of sensing elements because the sensors' usability is different regarding various applications. Currently, sensor array designers often perform application-oriented sensor selection. It is in principle against the idea of broad applicability. We propose another approach. It consists in that sensor selection does not refer to the physical presence of a sensor in an array but to the use of sensor output signal in gas assessment tasks, associated with particular application. The sensor array is composed of very many sensors. They all perform measurements. However, only selected combinations are involved in providing data for various pattern recognition tasks. In this way, a great flexibility regarding adjustment to numerous applications is reached.

While using commercial sensors as measuring elements in sensor array, the way to enhance the information content of sensor array response is to vary sensor operating conditions in course of exposure to a measured gas. In this respect, our original proposal is a stop-flow operation mode. It involves measurements in static and dynamic conditions, including both steady states of sensors and transients. The mode is simple in execution and it was shown to provide for a considerable increase in the number of best domains for pattern recognition.

A sensor array with broad applicability has to face numerous gas assessment problems of qualitative and quantitative kind. Any gas assessment performed by the sensor array is addressed at the data analysis level as pattern recognition problem. In currently offered arrays, different pattern recognition tasks are usually solved in one feature space and nonlinear pattern recognition methods are favoured. In our work we promote the idea of solving each gas assessment task using the feature space(s) where the associated pattern recognition problems are solved best. The best feature spaces are found in course of feature selection process using wrapper approach. The concept of extended sub-sampling is introduced as a means of generating numerous feature spaces where various factors associated with sensor array construction and operation are exploited to cause differences among features. Considering pattern recognition methods we retract from using complex ones in favour of fast learning classifiers and simple regression methods, preferably linear ones. Feature selection is to provide feature space(s) where these particular pattern recognition methods perform best. To improve the reliability of our results, many patterns which represent the same test gas are considered while performing test gas assessment. For that reason low dimensional feature spaces are favoured. Calibration of pattern recognition models, which operate in such spaces, may be satisfied using calibration data sets of modest size.

It is believed that the presented approach is original ant it will gain interest in the community of sensor array designers.

5. Acknowledgement

This work was supported by the project "Detectors and sensors for measuring factors hazardous to environment - modelling and monitoring of threats", POIG.01.03.01-02-002/08-00.

6. References

Amrani M.E.H., Payne P.A., Persaud K.C. (1992). Multi-frequency measurements of organic conducting polymers for sensing of gases and vapours. *Sens. Actuators B: Chem.*, Vol. 8, pp. 137-141

Arnold Ch., Harms M., Goschnick J. (2002). Air quality monitoring and fire detection with the Karlsruhe electronic micronose KAMINA. *IEEE Sensors Journal*, Vol. 2, No. 3, pp.179-188

Barbri N.E., Duran C., Brezmes J., Cañellas N., Ramírez J.L., Bouchikhi B., Llobet E. (2008). Selectivity Enhancement in Multisensor Systems Using Flow Modulation Techniques, *Sensors*, Vol. 8, pp. 7369-7379

Berna A. (2010). Metal Oxide Sensors for Electronic Noses and Their Application to Food Analysis. *Sensors*, Vol. 10, pp. 3882-3910

Bourgeois W., Romain A.C., Nicolas J., Stuetz R. M. (2003). The use of sensor arrays for environmental monitoring: interests and limitations. *J. Environ. Monit.*, vol. 5, pp. 852–860

Byun H.G., Persaud K. C., Pisanelli A.M. (2010). Wound-State Monitoring for Burn Patients Using E-Nose/SPME System. *ETRI Journal*. Vol. 32, No. 3 pp. 440-446

Faglia F., Comini E., Cristalli A., Sberveglieri G., Dori L. (1999). Very low power consumption micromachined CO sensors. *Sens. Actuators B: Chem.*, Vol. 55, pp. 140–146

Gardner J.W., Boilot P., Hines E.L. (2005). Enhancing electronic nose performance by sensor selection using a new integer-based genetic algorithm approach, *Sens. Actuators B: Chem.*, Vol. 106, pp. 114–121

Gouws G.J., Gouws D.J. (2003). Analyte identification using concentration modulation and wavelet analysis of QCM sensors. *Sens. Actuators B: Chem.*, Vol. 91, pp. 326–332.

Gualdr´on O., Llobet E., Brezmes J., Vilanova X., Correig X. (2006). Coupling fast variable selection methods to neural network-based classifiers: Application to multisensor systems. *Sens. Actuators B: Chem.*, Vol. 114, pp. 522–529

Gutierrez F.C., Ares L., Horrillo M.C., Sayago I., Agapito J.A., Lopez L. (1991). Use of complex impedance spectroscopy in chemical sensor characterization. *Sens. Actuators B: Chem.*, Vol. 4, pp. 359-363

Gutierrez F.C., Ares L., Robla J.I., Horrillo M.C., Sayago I., Agapito J.A. (1992). Properties of policrystalinne gas sensor based on d.c. and a.c. measurements. *Sens. Actuators B: Chem.*, Vol. 8, pp. 231-235

Gutierrez-Osuna R., Gutierrez-Galvez A., Powar N. (2003). Transient response analysis for temperature-modulated chemoresistors. *Sens. Actuators B: Chem.*, Vol. 93, pp. 57–66

Llobet E., Brezmes J., Vilanova X., Sueiras J., Correig X. (1997). Qualitative and quantitative analysis of volatile organic compounds using transient and steady-state responses of a thick-film tin oxide gas sensor array. *Sens. Actuators B: Chem.*, Vol. 41, pp. 13–22

Maciejewska M., Szczurek A., Ochromowicz Ł. (2009). The characteristics of a "stop–flow" mode of sensor array operation using data with the best classification performance. *Sens. Actuators B: Chem.*, Vol. 141, pp. 417–423

Maciejewska M., Szczurek A., Bodzoj L., Flisowska-Wiercik B. (2010). Sensor array and stop-flow mode applied to discrimination and quantification of gas mixtures. *Sens. Actuators B: Chem.*, Vol. 150, pp. 93–98

Pardo M., Sisk B.C., Sberveglieri G., Lewis N.S. (2006). Comparison of Fisher's linear discriminant to multilayer perceptron networks in the classification of vapors using sensors array data, *Sens. Actuators B: Chem.*, Vol. 115, pp. 647-655

Persaud K., Dodd G.H. (1982). Analysis of discrimination mechanisms of the mammalian olfactory system using a model nose. *Nature*, Vol. 299, pp. 352–355

Stetter J.R., Strathmann S., McEntegart C., Decastro M., Penrose W.R. (2000). New sensor arrays and sampling systems for a modular electronic nose. *Sens. Actuators B: Chem.*, Vol. 69, pp. 410–419

Szczurek A., Maciejewska M., Flisowska-Wiercik B., Bodzoj L. (2010). The stop-flow mode of operation applied to a single chemiresistor, *Sens. Actuators B: Chem.*, Vol. 148, pp. 522–530

Szczurek A., Maciejewska M., Flisowska-Wiercik B. (2011). Method of gas mixtures discrimination based on sensor array, temporal response and data driven approach. *Talanta*, Vol. 83, pp. 916–923

Tomchenko A.A., Harmer G.P., Marquis B.T., Allen J.W. (2003). Semiconducting metal oxide sensor array for the selective detection of combustion gases. Sens. Actuators B: Chem., Vol. 93, pp. 126–134

Vergara A., Martinelli E., Llobet E., Giannini F., D'Amico A., Di Natale C. (2006). An alternative global feature extraction of temperature modulated micro-hotplate gas sensors array using an energy vector approach. *Sens. Actuators B: Chem.*, Vol. 124, pp. 352–359

Vergara A., Llobet E., Martinelli E., Di Natale C., D'Amico A., Correig X. (2007). Feature extraction of metal oxide gas sensors using dynamic moments. *Sens. Actuators B: Chem.*, Vol. 122, pp. 219–226

Garrigues S., Talou T., Nesa D. (2004). Comparative study between gas sensors arrays device, sensory evaluation and GC/MS analysis for QC in automotive industry. *Sens. Actuators B: Chem.*, Vol.103, pp. 55–68

Zampolli S., Elmi I., Ahmed F., Passini M., Cardinali G.C., Nicoletti S., Dori L. (2004). An electronic nose based on solid state sensor arrays for low-cost indoor air quality monitoring applications. *Sens. Actuators B: Chem.*, Vol. 101, pp. 39–46

7

Neuro-Fuzzy Classifiers/Quantifiers for E-Nose Applications

Ravi Kumar
Thapar University, Patiala
India

1. Introduction

Smell is still a mystery to scientists in somehow, which cannot be studied with ease in vertebrates. Another problem is that the sense of smell is poorly developed in human beings in comparison with the same in many vertebrates (Menini *et al.*, 2004). This makes realization of an artificial olfactory system a challenging task. An artificial olfactory system (commonly known as E-nose) provides a low cost alternative to identification, quantification and characterization of odours. The traditional methods of characterization and quantification of odours generally involve the use of a trained panel of human experts. The use of human panel is sensitive to individual variability, adaptation (tendency to become less sensitive after prolonged exposure), mental state, fatigue, subjectivity, infections and exposure to hazardous compounds (Nagle *et al.*, 1998). Therefore, it is necessary to have a low cost and compact device to perform real-time analysis. It is thus natural for researchers to envisage a system, which is biologically inspired and modelled on the lines of an olfactory system. The increased understanding of the biological phenomenon of olfaction has motivated scientists to achieve artificial olfaction. Rapid strides made in the field of material science and fabrication technology has paved the way for manufacture of a large variety of micro-sensors, of which a large percentage is chemical sensors.

An E-nose uses multiple sensors in the form of an array. In an array of sensors each sensor responds broadly to a range or class of gases rather than a specific one. This characteristic of a sensor array is similar to a human nose, which is also partially sensitive to several odorants. The partial sensitivity of the sensor array can be exploited for characterization and quantization of gases/odours by making use of effective signal processing and pattern recognition. In an electronic nose, the odorants produce changes in physical/chemical properties. A sensor array converts the chemical inputs into electrical signals which are further processed by utilizing an electronic circuit, providing an analogue signal to be amplified, pre-processed and/or digitised prior to being fed into a pattern recognition system (Shurmer *et al.*, 1990; Nakamoto *et al.*, 1990). Basic stages of an artificial olfactory system are shown in Fig. 1.

After coming in contact with the odorants, the sensors experience a change in electrical properties. Each sensor is sensitive to all the odorant molecules in their specific way. Most electronic noses use sensor arrays that react to gases/odours on contact: the adsorption of gases/odours on the sensor surface causes a change in physical properties of the sensor (Sarro, 1992). A specific response is recorded by the electronic interface transforming the signal into a digital value. Recorded data are then analyzed using computational techniques (Osuna *et al.,* 2002). An artificial olfaction system can be fabricated using standard micro-electronic techniques for on-chip integration. An E-nose employs an array of chemical sensors to achieve an appreciable level of selectivity to different gases/odours. Metal oxide based sensors fabricated with the thick film technology are the most popular choice for gas/odour sensing. Most of the commercial E-noses employ metal oxide based sensing devices because metal oxides are most suited as gas sensors due to their high sensitivity and their ability to maintain structural integrity in harsh conditions, namely, high temperature (Moseley, 1992). The basic sensing mechanism in metal oxide based sensors involves a change in resistance due to chemisorptions when exposed to odorants/gases.

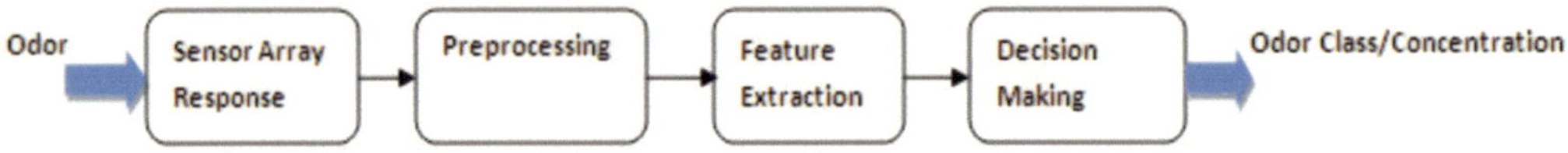

Fig. 1. Artificial olfactory system

1.1 Operating principle of tin oxide gas sensors

It has been found that when a bead of tin oxide is heated in the presence of a combustible contaminant, and the conductance of the bead is measured continuously, it is possible to obtain a measure of the concentration of the contaminant gas (Watson, 1984). This observation can be explained as follows.

By heating a bead of tin oxide in clean air, oxygen can be adsorbed onto the surface layers until equilibrium is achieved for that particular temperature. The measurement of the characteristic conductance of the bead would reveal that it is a function of both the temperature and partial pressure of the oxygen. Significant change in surface conductivity of semiconductors can be brought about by adsorption and subsequent reaction of gases with the adsorbed oxygen. The active material of the sensor is generally SnO_2, which is an n-type semiconductor. When oxygen is adsorbed on to it, it accepts electrons to become O_2^- O^- or O^{2-} (Ikohura, 1981). The adsorption of a reducing gas releases bound electrons and thus increases the conductance of the surface dramatically. For an oxidizing gas converse mechanism operates. Various dopants are used to improve the sensitivity and selectivity of thick film tin oxide gas sensors (Morrison, 1987).

Sensor operating temperature plays a vital role in the development of gas selective sensors. Since, different classes of reducing gases have different reaction rates, sensors operating at different temperatures show a degree of selectivity (Sears *et al.,* 1990). Despite having appreciable sensitivity to a large number of gases/odours, thick film tin oxide sensors have some well known limitations such as, cross sensitivity to a number of compounds and saturation of sensor response at higher concentration of the odorants. These limitations can be overcome by employing an array of sensors whose responses are analyzed subsequently using appropriate pattern analysis techniques. Typical materials used for fabrication of the sensors are SnO_2, ZnO, Fe_2O_3, and WO_3. These metal oxide films are used with or without dopants like CuO, Pd, Pt and In to enhance their selectivity. Porous and sintered SnO_2 is the most widely used for gas sensors, as it is appreciably sensitive to a large number of gases/odours. SnO_2 sensors are available both for domestic and commercial use. For domestic applications, devices are available for detecting combustible gases such as CO, H_2, alcohols, LPG, and volatile matters from food stuff. For industrial applications detectors are available for gases such as NH_3, H_2, H_2S, CH_4, C_7H_8, C_8H_{10} and hydrocarbons. The following subsection presents a typical experimental set up where a sensor array is exposed to several odorants and the response of each sensor is noted. This particular set up is chosen for illustration because its response pattern is the most challenging from pre-processing and computational point of view. The same data would be used throughout the chapter for demonstrating the efficacy of the computational techniques employed. Published data from other sensor arrays may also be reproduced to explain some of the computational challenges.

1.2 Typical experimental set up for odour sensing

Integrated gas sensor array comprises several sensors as shown in Fig. 2. Sensors are fabricated on one side of an alumina substrate whereas a resistive pattern is fabricated on the other side to achieve uniform heating. A metal oxide paste is prepared, which is applied to the substrate and fired at high temperature by passing it to a furnace so that the paste sticks properly to the substrate. Different dopants (e.g. ZnO, Sb_2O_3 and NiO) are used respectively with the metal oxide paste, resulting in different types of gas sensors with different sensitivity to different odours. The diagram of a tin oxide sensor array pattern is shown in Fig. 2. The fabricated integrated sensor array pattern is then tested under closely controlled environmental conditions using some experimental chamber with a facility (either manual or automated) of injecting test gases as shown in Fig. 3. The experiment is designed for testing 4 types of whiskies, two types of rums, and ethanol (Nayak *et al.* 1992).

Initially, the sensor array is kept in a closed ambient air under energized condition at 10W heater supply for more than 30 minutes to make the sensor resistances stable. At this stage, initial resistance of the sensors is recorded. Then a drop of test alcohol is injected into the chamber and it gets vaporized to gaseous phase before being adsorbed on to the sensor surface. Sensor readings are reported after three minutes as it is found to be optimum time to equilibrate with the sample alcohol. Similarly, another drop of test alcohol is injected and the experiment is repeated for more drops so that the observations are made for up to 12 drops of test alcohols. After experimenting with one of the alcohols/alcoholic beverages, the sensors are recovered in open ambient air at room temperature and on complete recovery; experiments are repeated for other alcohols/alcoholic beverages. Using suitable mathematical manipulations, the concentration of test odorants can be converted to parts per million (ppm).

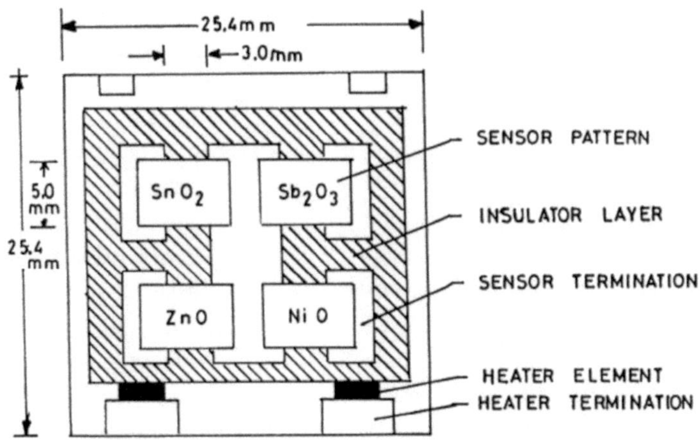

Fig. 2. Fabricated integrated sensor array pattern (Nayak *et al.*, 1992)

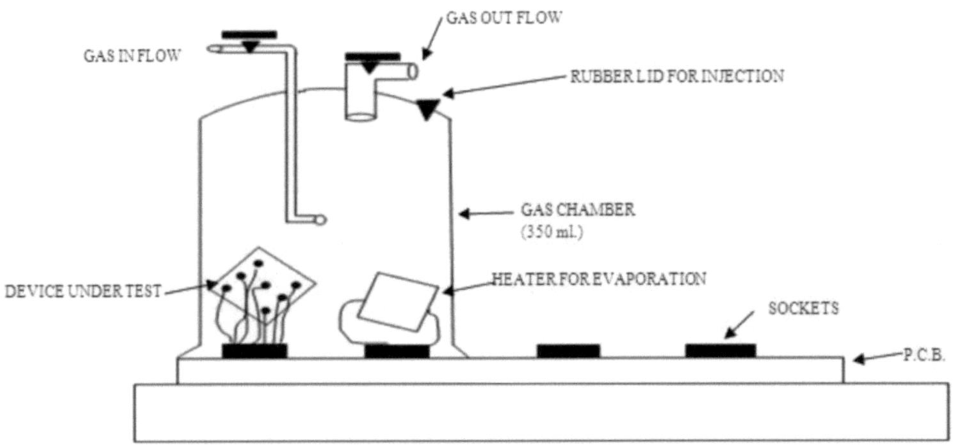

Fig. 3. Experimental chamber for exposing sensor array to gases/odours

The next step is to obtain the sensor response by calculating the percentage change in resistance of all the sensors for all the odorants injected into the test chamber. This is done to nullify the effect of initial resistances. The percentage change in resistance is calculated using

$$P_{ijd} = \frac{R_{ijo} - R_{ijd}}{R_{ijo}} \leq 100 \tag{1}$$

where, R_{ijo} is the initial resistance of the ith sensor for zero-th drop of jth odorant and the subscript d denotes a particular drop. The steady-state exposure profiles of the sensor array exposed to different types of alcohols and alcoholic beverages at different concentrations thus obtained is shown in Fig. 4.

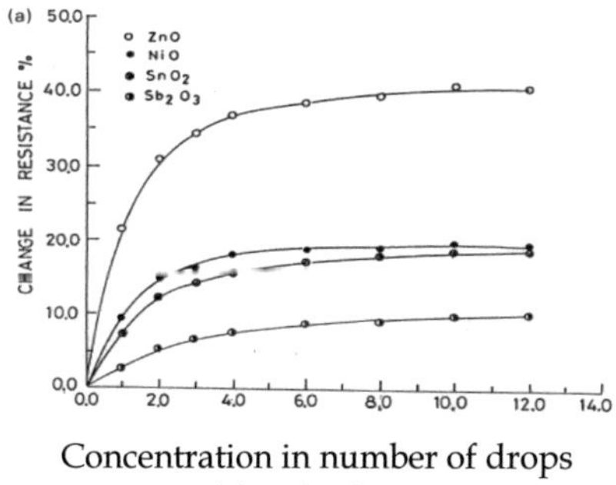

(a) Whisky-1

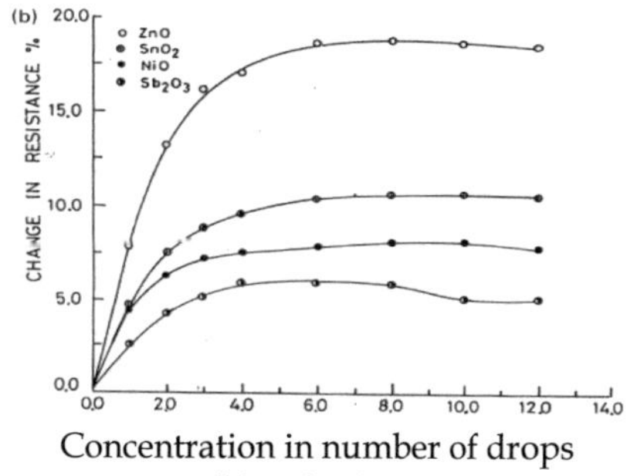

(b) Whisky-2

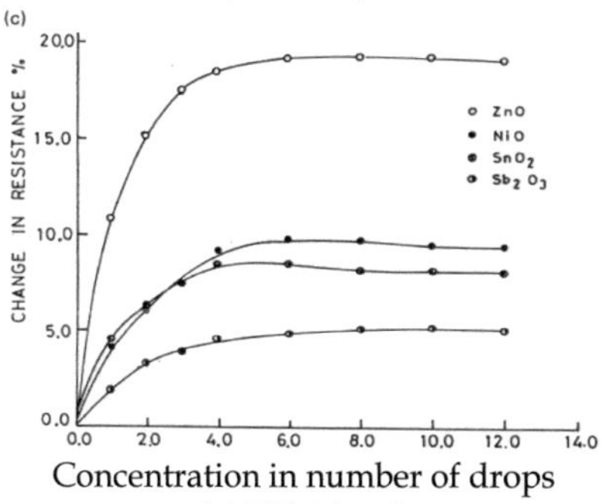

(c) Whisky-3

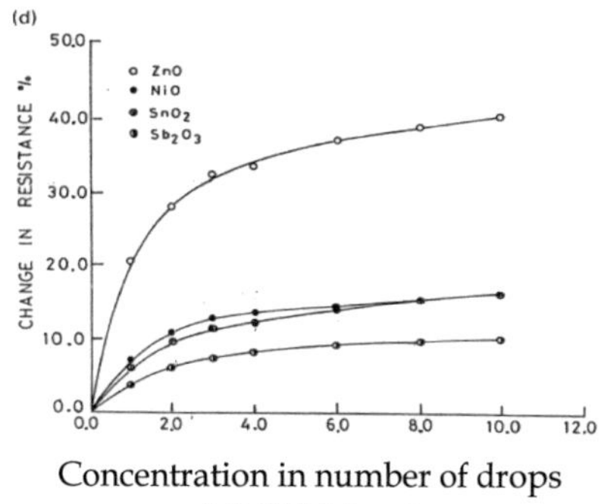

(d) Whisky-4

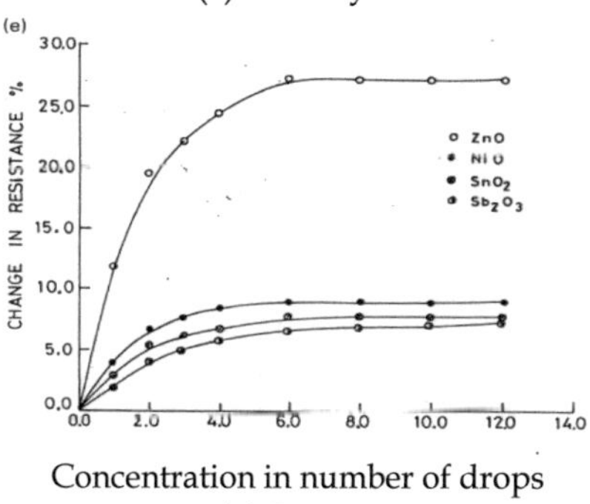

(e) Rum-1

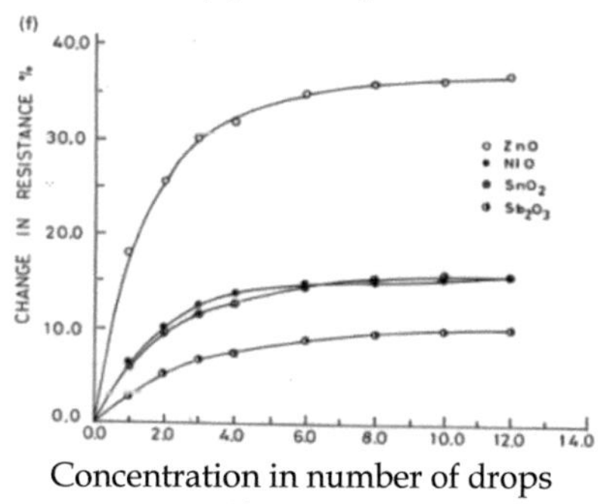

(f) Rum-2

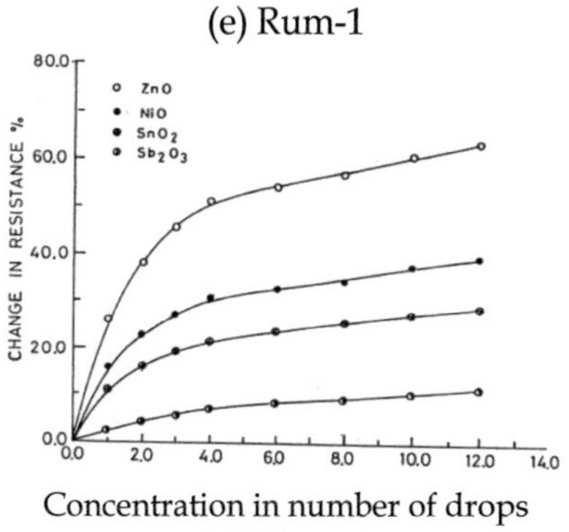

(g) Ethanol

Fig. 4. Steady-state response of sensor array upon exposure to different alcoholic beverages (Nayak *et al.*, 1992)

1.3 Limitations of sensor arrays

In general, sensor arrays suffer from one or more of the following limitations:

1. Overlapping sensitivity to different gases/odours leading to poor selectivity.
2. Saturating tendency of the sensor response at higher concentration of the test gas leading to difficulties in quantification
3. "Drift" in the sensor response, which is defined as the variation in the output of a sensor when exposed to a particular test gas under identical conditions after a finite interval of time.

1.3.1 Overlapping sensitivity

Overlapping sensitivity is by far the most challenging limitation of a sensor array. It renders a nicely fabricated sensor array less capable of discriminating between two gases/odours in spite of having an appreciable sensitivity.

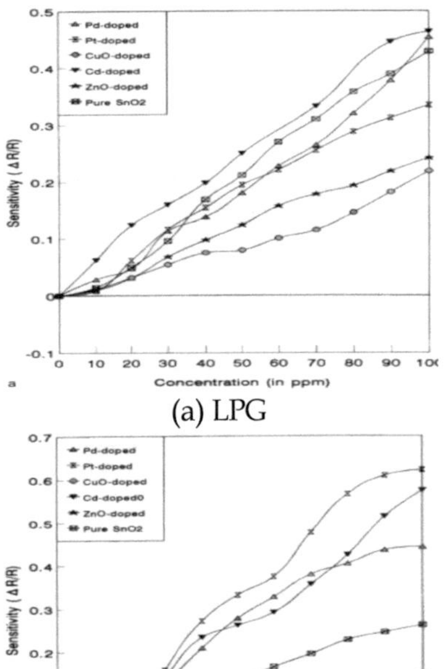

(a) LPG

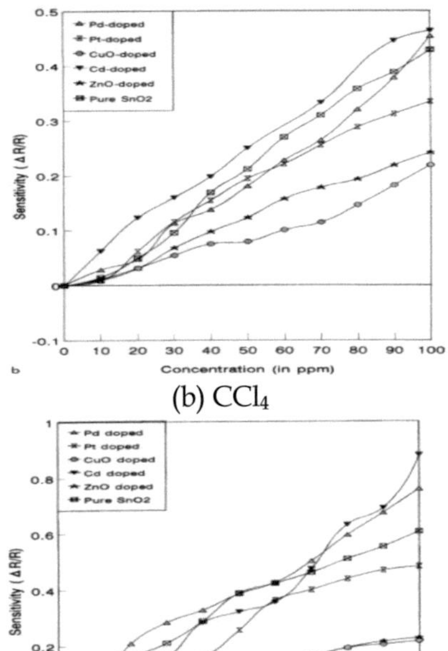

(b) CCl_4

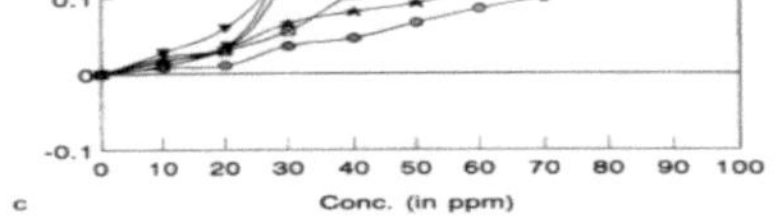

(c) CO

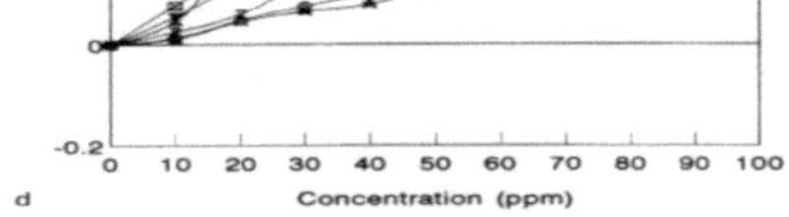

(d) C_3H_7OH

Fig. 5. Sensitivity characteristics of oxygen plasma-treated array at different concentrations (Chaturvedi *et al.*, 1999)

Fig. 5 shows the plot of sensitivity with concentration of the test gas for a 6-sensor array. It is clear from Figs. 5 (a) and (b) that the sensor array exhibits almost identical sensitivity for LPG and CCl_4 and hence representing a very poor selectivity for these gases. It should be noted that all the sensors of the array are appreciably sensitive to all the 4 test gases. In spite of this fact we would not be able to discriminate between two of the 4 gases (i.e. LPG and

CCl_4). Therefore, in addition to having good sensitivity, the sensors in the array should respond differently to different test gases/odours.

1.3.2 Saturation and drift

Fig. 4 shows the response vs concentration of an array of 4 sensors exposed to the vapour of a particular alcoholic beverage. It can be seen that responses of almost all the sensors saturate after a particular concentration of test gas has been injected into the experimental chamber. This phenomenon makes the quantification of the test gas impossible at higher concentrations.

Drift introduces an unwanted temporal variation in the sensitivity of a sensor array. This means the response of the array to same gases under identical conditions may be entirely different from what was obtained previously. When previously learned sensor patterns become obsolete, the ability of the sensor to discriminate is lost. In fact, sensor drift is the highest obstacle in the wide marketability of low cost gas sensors.

All the above mentioned problems are hindrance to proper identification and quantification of gases/odours. Thick film sensors are well known for their design ruggedness, ease of fabrication, sensitivity to a plethora of gases/odours, and most importantly for being economical. The above mentioned limitations of thick film sensors negate their other desirable features discussed above and hence, a promising technology sometimes seems to fall short of achieving its objectives. The limitations imposed by poor selectivity and response saturation can be overcome by employing computational techniques to extract both qualitative and quantitative information. The role of computational techniques in gas/odour discrimination can never be underestimated. Wherever possible, putting more emphasis on computational methods can save a significant amount of resources since breaking innovation in the fabrication technology requires time and effort. The next section presents an overview of computational challenges put forth by popular sensors with an aim of proper identification and quantification of individual odorants.

2. Computational challenges

Choice of an appropriate technique is highly dependent on the problem in hand. In the context of E-nose systems, the term pattern analysis applies to both qualitative and quantitative analysis of odours. The response data generated from a sensor array are multivariate in nature. There are several issues, which require careful consideration for a successful design of a pattern analysis system. Signal pre-processing, feature extraction, feature selection, classification, regression, clustering and multi-fold cross-validation are the most prominent goals of a pattern analysis system, for which critical design issues are to be taken care of. The first computational stage in a pattern analyzer is often the signal pre-processing stage. The main purpose of a pre-processing stage is to select a number of parameters that are descriptive of the sensor array response. The choice of parameters can significantly affect the performance of the subsequent modules in the pattern analysis system. Fig. 6 (a) shows the scatter plot for S-1 as SnO_2 and S-2 as SnO_2 doped with Sb_2O_3 for the 4 sensor array mentioned in the previous section.

The plots confirm that the clusters are not only overlapping but also a high degree of scattering of data points. The overlapping of clusters is due to the cross sensitivity and is attributed to the material properties of the sensors. The goal of a pre-processing stage is to minimize the spread in an individual clusters and maximize the distance between two clusters. Therefore, a pre-processing technique should be applied, which utilizes the statistical properties of the data set to maximize their inter class separation and minimize the intra class separation. The result of such a pre-processing is shown in Fig. 6 (b) to establish the importance of a pre-processing stage. The technique used is Transformed Cluster Analysis (TCA). Details of TCA can be read from a published work of the author (Kumar *et al.* 2010).

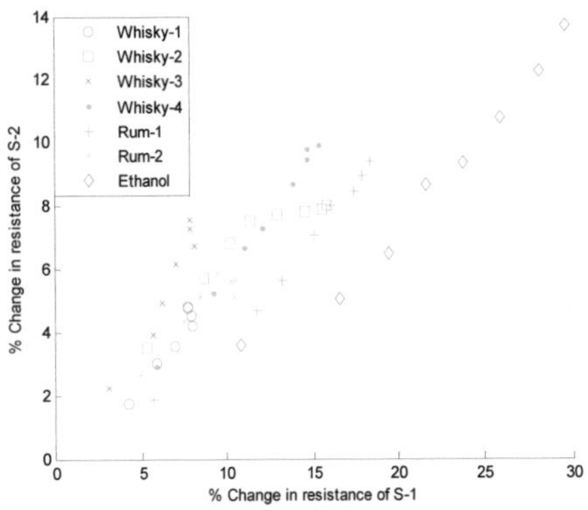

(a) 2-D scatter plot for responses of sensors S-1 and S-2

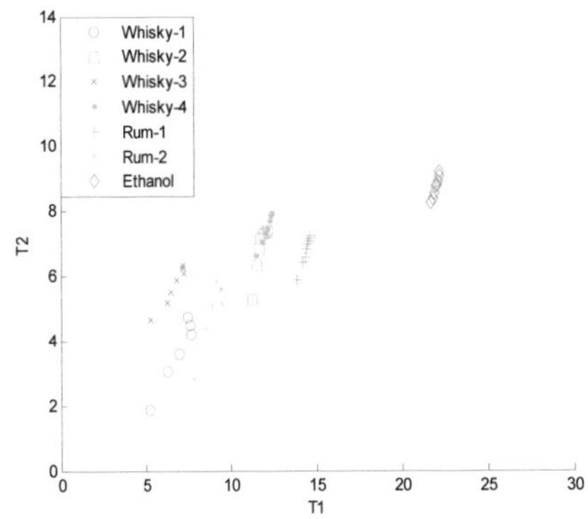

(b) 2-D scatter plot for transformed responses of sensors S-1 and S-2 (T1 and T2 are transformed responses of sensors S-1 and S-2)

Fig. 6. 2-D scatter plot for responses and transformed responses of sensors S-1 and S-2

The actual identification/quantification part of pattern analysis begins after pre-processing. Pattern analysis techniques are generally of two types viz. parametric and non-parametric techniques. Parametric techniques do not require any prior information on the type and number of different classes contained in data. In non-parametric techniques a set of response patterns is compared against each other on the basis of degree of similarity or dissimilarity (Gardner, 1987). Thus, non-parametric techniques are more general in nature.

Statistical pattern analysis techniques like Principal Component Analysis (PCA) and Cluster Analysis (CA) are one of the most popular non-parametric techniques. PCA overcomes the "curse of dimensionality" introduced by the response vector of a multi-sensor array by choosing "principal components" along the directions of maximum variance. Principal components are a linear combination of original variables with the redundant information eliminated. The reduced dimensionality of data makes the subsequent feature extraction task simpler.

Feature extraction attempts to find a low dimensional mapping that preserves most of the information in the original feature vector. The mappings thus formed enhance the information content of the feature vector. Feature extraction techniques also help signal representation, which can be useful for extrapolatory data analysis. They are helpful in visualizing high dimensional data. Most of the feature extraction techniques for E-nose applications are based on PCA, which is a signal representation technique that generates projections along the directions of maximum variance. Learning in pattern analyzers is viewed as the optimization of a process to obtain a minimum value for a solution of a pre-specified objective function (criterion). Analysis of patterns by an analyzer is carried out either by supervised or unsupervised learning schemes.

The pattern analysis techniques applied to the output of a sensor array should be biologically inspired if an E-nose is to sniff like humans. This requires application of biologically inspired algorithms to the senor output. Artificial neural networks (ANNs) are such a class of computational paradigms, the inspiration for which comes originally from the studies of mechanism of information processing in biological nervous system, particularly brain (Bishop, 1994). The advantages of ANNs include massive parallelism, distributed processing and computation, learning ability, generalization ability and adaptability. Apart from ANNs, fuzzy logic and genetic algorithm are some other techniques which constitute a class of paradigms known as "soft computing". Soft computing is fast replacing statistical learning techniques in pattern analysis applications. Also, a lot of work has been done in the area of gas/odour discrimination using soft computational techniques. Most of the techniques described above have been adopted from the field of chemometrics and they can be labelled as 'statistical' pattern recognition techniques as opposed to soft computational techniques, which are recent in origin and are in general biologically inspired. "Soft Computation" is a name given to a class of computational paradigms, which seek to find approximate solutions to ill-posed problems. It is tolerant of imprecision, uncertainty, partial truth, and approximation just like the human mind. The principal constituents of soft computing are ANNs, Fuzzy Logic, Support Vector Machines, and Evolutionary Computing.

Soft computational techniques have revolutionized the arena of artificial olfaction by immensely reducing dependency on flawless and meticulously designed sensor hardware. Proper application of soft computational techniques can improve the discrimination obtained using the response of poorly selective sensors to a great extent thus, saving additional costs on possible replacement and fabrications of novel sensor hardware. ANNs are one of the foundation pillars of soft computing. Their enormous learning capability, massively parallel architecture, and availability of a large number of learning algorithms for their training makes them a popular choice for a wide variety of computational tasks. A close scrutiny of the available literature reveals that the choice of pattern analysis techniques for artificial olfaction has been highly problem-dependent (Osuna, 2002). Different types of sensor arrays generate response data with different statistical properties making selection of an appropriate technique a difficult task. In the context of ANNs, the choice is often between a lesser architectural complexity and a lower system error. In view of all these, in the next section identification task of 7 different alcohols and alcoholic beverages is taken up using ANN and the response of the 4 sensor array described in section one.

3. ANNs for odour identification

The basic computational unit in an ANN is neuron, which is a mathematical function used to approximate input-output mappings. The output of a neuron (also known as firing of a neuron) is dependent upon the synaptic weight connection between all the neurons in the network. This synaptic weight ensemble changes if the actual output is not equal to the desired output. The process of change in synaptic weight is known as 'learning' or 'training' of ANN. A well trained ANN can perform any task related to classification, function approximation or prediction with some small amount of error with a brand new data set. In ANNs neurons are arranged in different layers with 'hidden layer' being responsible for performing most of the computational tasks. ANNs can be trained to approximate any non-linear input-output mapping. It can be shown mathematically that any ANN trained with orthogonal least squares algorithm is able to approximate any input-output mapping with arbitrary accuracy provided the number of neurons in the hidden layer is large enough (Hykin, 2009). However, a back-propagation (BP) algorithm is by far the most popular method of training an ANN. It is less complex and requires lesser number of neurons to perform the computations.

3.1 Back-propagation algorithm

BP is an algorithm where input vectors and the corresponding target vectors are used to train a network until it can approximate a function, associate input vectors with specific output vectors, or classify input vectors in an appropriate way as defined by the user. Networks with biases, a hidden layer, and a linear output layer are capable of approximating any function with arbitrary accuracy. A standard BP is a gradient descent algorithm, in which the network weights are moved along the negative of the gradient of the performance function. The term *BP* refers to the manner, in which the gradient is computed for nonlinear multilayer networks. Basically, error BP consists of two passes through the different layers of the network: (1) forward pass, and (2) backward pass.

In the forward pass the input vectors are applied to the sensory nodes of the network, and its effect propagates through the network layer-by-layer. The actual response of the network is delivered by the output nodes in the form of an output vector. The outputs are compared with a target vector and the difference is generated as error. Let the error signal at the output of neuron j at iteration n be defined by:

$$e_j(n) = d_j(n) - y_j(n) \tag{2}$$

where, d_j (n) represents the desired output at the output node j at iteration n, and $y_j(n)$ be the actual output at the output node j at iteration n.

Let the instantaneous value of the error energy for neuron j be defined as:

$$\frac{1}{2}e_j^2(n)$$

Then, for all neurons in the output layer instantaneous value ξ (n) of the total energy is given by:

$$\xi(n) = \frac{1}{2}\sum_{j \in C} e_j^2(n) \tag{3}$$

where, C is the set of all neurons in the output layer of the network. Let N denote the total number of patterns contained in the training set. The average squared energy over the entire training sample is now given by:

$$\xi_{avg}(N) = \frac{1}{N}\sum_{n=1}^{N}\xi(n) \tag{4}$$

For a given training set, ξ_{avg} is called the cost function, which is a measure of learning performance. Minimization of the cost function is done iteratively. The weights associated with the network are updated on a pattern-by-pattern basis until one complete presentation of the entire training set (epochs) has been done. The adjustments to the weights are made in accordance with the respective errors computed for each pattern presented to the network. The arithmetic average of these individual weight changes over the entire training sets presents an estimate of the true change that would result from modifying the weights based on minimizing the cost function ξ_{avg} over the entire training set. Fig. 7 shows a neuron *j* being fed by a set of input signals produced by a layer of its neurons to its left.

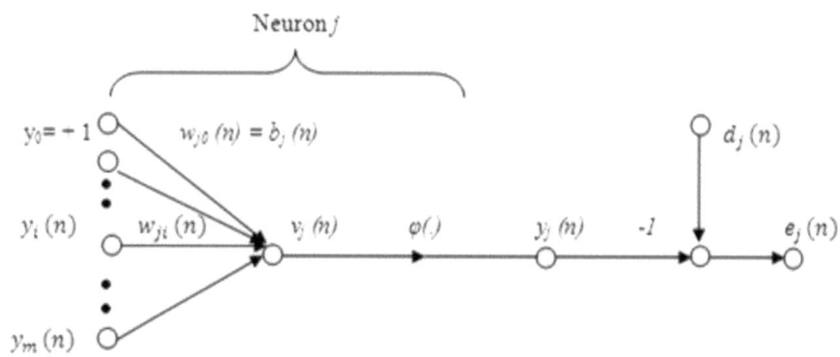

Fig. 7. Neuron 'j' being fed by a set of signals from a previous layer of neurons

The induced local field $v_j(n)$ produced at the input of the activation function associated with neuron *j* is given by,

$$v_j(n) = \sum_{j=0}^{m}\omega_{ji}(n)y_i(n) \tag{5}$$

where m is the total number of inputs applied to neuron *j*, and ω_{ji} is the synaptic weight from neuron *i* to neuron *j*. The signal $y_j(n)$ appearing at the output of neuron *j* at iteration n is a function of the induced local field

$$y_j(n) = \varphi\{v_j(n)\} \tag{6}$$

The BP algorithm applies a correction $\Delta\omega_{ji}(n)$, to the synaptic weight $\omega_{ji}(n)$, which is proportional to the partial derivative

$$\frac{\partial\xi(n)}{\partial w_{ji}(n)}$$

Applying the chain rule of calculus, this gradient can be expressed as:

$$\frac{\partial \xi(n)}{\partial \omega_{ji}(n)} = \frac{\partial \xi(n)}{\partial e_j(n)} \frac{\partial e_j(n)}{\partial y_j(n)} \frac{\partial y_j(n)}{\partial v_j(n)} \frac{\partial v_j(n)}{\partial \omega_{ji}(n)} \tag{7}$$

The partial derivative

$$\frac{\partial \xi(n)}{\partial w_{ji}(n)}$$

represents a sensitivity factor. It determines the direction of search in weight space for the synaptic weight ω_{ji}.

Differentiating both sides with respect to $e_j(n)$ we get

$$\frac{\partial \xi(n)}{\partial e_j(n)} = e_j(n) \tag{8}$$

Differentiating both sides of (2) with respect to $y_j(n)$, we get,

$$\frac{\partial e_j(n)}{\partial y_j(n)} = -1 \tag{9}$$

Differentiating both sides of (6) with respect to $v_j(n)$, we get,

$$\frac{\partial y_j(n)}{\partial v_j(n)} = \varphi_j'\{v_j(n)\} \tag{10}$$

Also, differentiating (5) with respect to $\omega_j(n)$, we get,

$$\frac{\partial v_j(n)}{\partial w_{ji}(n)} = y_i(n) \tag{11}$$

The use of (8) to (11) in (10) yields:

$$\frac{\partial \xi(n)}{\partial \omega_{ji}(n)} = -e_j(n)\,\varphi_j'\{v_j(n)\}y_i(n) \tag{12}$$

The correction $\Delta\omega_{ji}(n)$ applied to $\omega_{ji}(n)$ is defined by the delta rule:

$$\Delta\omega_{ji}(n) = -\eta \frac{\partial \xi(n)}{\partial \omega_{ji}(n)} \tag{13}$$

where, η is the learning-rate parameter of the BP algorithm. The gradient descent in weight space takes place in a direction for weight change that reduces the value of $\xi(n)$The use of (12) in (13) yields

$$\Delta w_{ji}(n) = \eta \delta_j(n) y_i(n) \tag{14}$$

where, the local gradient $\delta_j(n)$ is defined by

$$\delta_j(n) = -\frac{\partial\xi(n)}{\partial v_j(n)} = -\frac{\partial\xi(n)}{\partial e_j(n)}\frac{\partial e_j(n)}{\partial y_j(n)}\frac{\partial y_j(n)}{\partial v_j(n)} \tag{15}$$
$$= e_j(n)\varphi_j'\{v_j(n)\}$$

The local gradient points to required change in synaptic weights. Hence, the above relation between the learning rate, local gradient and weight correction can be summarized as follows:

(Weight Correction) = (Learning Rate)(Local Gradient)*(Input Signal of Neuron 'j')*

Learning Rate Parameter and Momentum Constant:

The learning rate parameter η is a measure of the changes to the synaptic weights in the network over subsequent iterations. Thus, a smaller learning rate parameter makes smaller changes in synaptic weights and the trajectory in the weight space is smoother. A smaller learning rate results in a slow learning. If the learning rate parameter η is made too large to speed up the learning process, the resulting large changes in the synaptic weights assume such a form that the network may become unstable. To avoid the danger of instability while keeping the learning rate fast enough, another term is added to the delta rule, which is known as the momentum constant and is denoted by α. Hence, Eq. 14 becomes:

$$\Delta w_{ji}(n) = \alpha \Delta w_{ji}(n-1) + \eta \delta_j(n) y_i(n) \tag{16}$$

The inclusion of momentum term in the BP algorithm has a stabilizing effect in directions that oscillate in sign. The momentum term also prevents the learning process from terminating in a shallow local minimum on the error surface. The training through a BP algorithm proceeds iteratively. A prescribed set of training examples are fed repeatedly to the ANN. The learning process continues on an epoch-by-epoch basis till the stabilization of the synaptic weights and bias levels of the network, and most importantly, the convergence of the average squared error over the entire training set to some minimum value. A back-propagation algorithm cannot converge. However, it is considered to have converged when the absolute rate of change in average squared error per epoch is "sufficiently small". The rate of change in average squared error is typically considered to be small enough if it lies in the range of 0.1 to 1% per epoch.

The sensor response curves of Fig. 4 were sampled at equal intervals of concentration and a data set was prepared. This data set will now be used to train an ANN using a BP algorithm.

3.1.1 Identification with sampled data

A three layer feed-forward neural network with sigmoidal activation function was simulated for the classification task. Neural network simulation was implemented in MATLAB using TRAINGDM function. The number of neurons in the input and output layers were fixed as 4 and 7 respectively as there are 4 sensors and 7 classes of gas/odour. Simulated network was trained by input vectors available with training data set while learning parameters such as learning rate (η) and momentum constant (α) and the number of neurons in the hidden layer were optimized during experimentations. After several repeated experiments the optimum number of neurons in the hidden layer was found to be six as it gave the minimum system error, which is measured in terms of mean square error (MSE) .The optimized BP network with a configuration of 4:6:7 and an optimized set of weights and biases was trained repeatedly with 4 sets of training data and by changing the values of learning rate (η) and momentum constant (α) from 0.1 to 0.9. The network was trained for a fixed 30,000 epochs with an error goal of zero. The trained network for minimum system error was then tested with 4 different test data sets. The training and test performance for all values of learning rate and momentum constants was noted. System error was studied at a particular learning rate for different values of the momentum constants in the range of 0.1 to 0.9 for 4 different testing subsets.

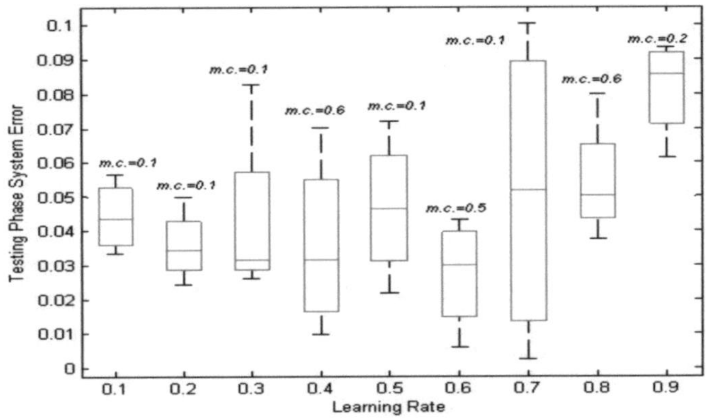

Fig. 8. Box whisker diagram for testing phase system error with learning rate for 4 testing subsets for different values of the momentum constant *(m.c.)* (BPNN)

Fig. 8 shows the testing phase system error at a particular value of learning rate and a momentum constant corresponding to which minimum average system error is observed. The BP neural network (BPNN) trained with raw data exhibited poor classification performance for all training subsets and also very high system error in both training and testing phases for almost all combinations of learning rates and momentum constants. A high degree of spread in system error is visible from Fig. 8 implying the inconsistent performance of BPNN trained with raw data. The diagrams show the variation in error performance with an optimum combination of learning rate and momentum constant calculated over 4 trials with different testing subsets.

3.2 Identification performance of radial basis function neural network

Apart from being subjected to a BP-trained neural classifier, the sampled data were fed as inputs to a Radial Basis Function Neural Network (RBFNN) classifier. RBFNN was chosen as a classifier because it takes shorter time to train apart form having a lower system error. In the following subsection, a brief introduction to RBFNNs is given and the subsequent subsections describe the identification performance of the RBFNN with respect to the present problem.

3.2.1 Radial basis function neural network

The radial basis function network is primarily composed of three layers. The first is an input layer. The second layer is the hidden layer which is primarily responsible for computation and the third layer consists of neurons with linear activation functions and it provides the output of the network corresponding to the input patterns (Hykin, 2009). The training of RBFNNs involves providing the best fit to the training data by finding a surface in a multidimensional space. An interpolation between the data points is performed in the testing phase. The RBFNN solves a classification problem by applying a nonlinear transformation from input space of lower dimension to the hidden space of higher dimension, since it increases the likelihood of correct classification for the given problem (Cover, 1965). The most popular learning strategy of RBFNNs involves the use of Gaussian functions with the selection of centres being done in a random manner. The standard deviation of the Gaussian function is fixed according to the "spread" of the centres. Given below is a radial basis function, with centre at '*t*'

$$G(||x-t_i||^2)\exp(-\frac{m_1}{d_{\max}^2}||x-t_i||^2) \quad i=1,2,.....m \tag{17}$$

where, m is the number of centres and $d_{\max}$ is the maximum distance between the randomly chosen centres. Also, the standard deviation (spread) of the RBF is given by

$$\sigma=\frac{d_{\max}}{\sqrt{2m_1}} \tag{18}$$

Thus, the learning process undertaken by RBFNN involves the optimization of the hidden layer's activation functions and the optimization of the output layer's weights. Fig. 9 shows a typical radial basis neuron. The net input to the radial basis transfer function is the vector distance between its weight vector and input vector multiplied by the bias, which allows the sensitivity of the neuron to be adjusted. The equations used in the neural model are given by:

$$o=\|w-p\|.b \tag{19}$$

$$\text{b}=\frac{0.833}{s} \tag{20}$$

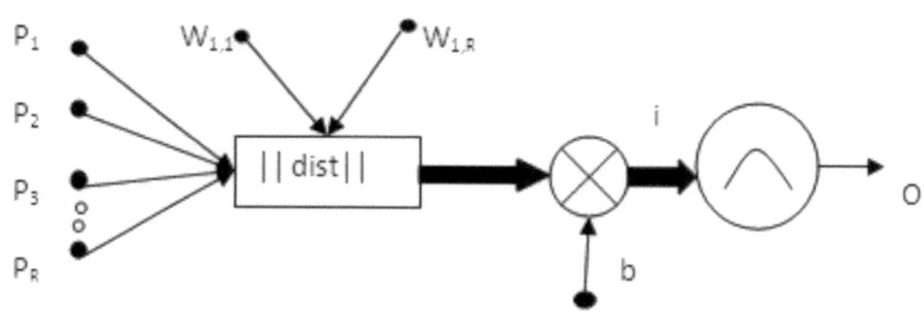

Fig. 9. Radial basis function neuron

where, o and b denote the output of a typical radial basis neuron and its bias, respectively.

The quantity is known as spread constant and is the most important learning parameter of a radial basis network. The radial basis function has a maximum of 1 when its input is 0. As the distance between w and p decreases, the output increases. Thus, a radial basis neuron acts as a detector that produces 1 whenever the input p is identical to its weight vector w. As is evident from the plot of radial basis function the function returns a value of 0.5 when the net input to radial basis transfer function is 0.833. The bias is given by Eq. (20). This determines the width of an area in the input space, to which each neuron responds. The spread constant should be large enough so that neurons respond strongly to overlapping areas of the input space.

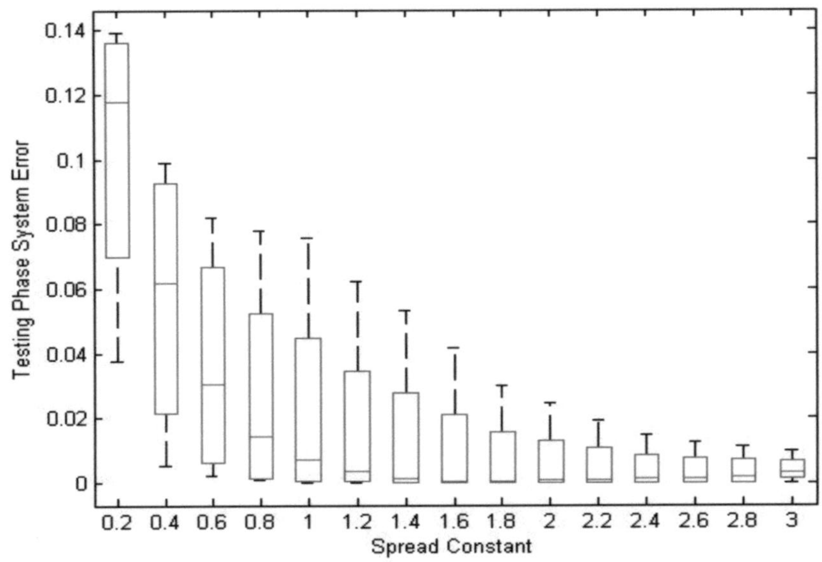

Fig. 10. Box whisker diagram for testing phase system error with spread constant

3.2.2 Identification results of RBFNN classifier

RBFNN employed in this study utilizes the *newrbe* function implemented in MATLAB. The function creates a radial basis network with the number of neurons in the hidden layer equal to the number of training patterns. The network was simulated first with the training data sets and was tested with test data sets. The spread constant of the network was varied from the 0.2 to 3.0 at regular intervals. The results thus obtained have been depicted in the form of a box whisker diagram of Fig. 10. K-fold cross validation scheme was used with K=6, to avoid overfitting. It is evident from Fig. 10 that as the spread constant increases, the variation in the results decreases. The minimum testing phase system error was obtained at a spread constant of 2.6 and 100% identification was achieved.

4. Fuzzy sets for odour discrimination

The foundations of fuzzy logic are based on the concept of fuzzy sets. A fuzzy set is a set without a clearly defined boundary (Zadeh, 1965). Human smell processing is inherently fuzzy in nature. When the qualitative remarks are used about something, we actually do a fuzzy classification task, in which our sensory responses are assigned to more than one predefined classes with varying degrees of belongingness to them. This *degree of belongingness* is known as degree of membership in fuzzy set theory. The qualitative remarks come in the form of linguistic labels such as *rose-like, apple-like*. Apart from this, human olfactory system is capable of doing multi-way classification. Given 3 types of fragrances to smell, a human being is able to tell, which one was *apple-like* or *rum-like* or *rose-like* and also that which fragrance among the three was strongest, which one was weakest and which one was of in-between intensity. In the above case, along with the qualitative information some quantitative information has also been retrieved, which enables us to label the fragrances according to their 'intensity'. This has served as the primary motivation for the design of a network, which can retrieve both the qualitative and quantitative information when the sensor array response vectors are given as input vectors to the network.

Fuzzy set theory is a generalization of the conventional crisp set theory. It measures the degree to which an event occurs (Zadeh, 1965). As discussed above, each element of a fuzzy set has a degree of membership assigned to it in accordance with a membership function. The most commonly used membership functions in the literature being *triangular* and *trapezoidal* membership functions.

Let $X = \{x_1, x_2, x_3, \ldots\ldots, x_n\}$ be a non-fuzzy set. The subsets of X are called bit vectors or bivalent messages. If $X = \{x_1, x_2, x_3, x_4\}$, then $X = \{1,1,1,1\}$, $\varphi = (0,0,0,0)$ and the subset $A = \{x_1, x_4\}$ is represented as $A = (1, 0, 0, 1)$. The 1s and 0s indicate the presence or absence of the ith element x_i in the subset. Each non fuzzy subset A can be defined as one of the two-valued membership functions $\mu_A : X \rightarrow \{0,1\}$. The power set 2^X of X is the set of all of X's subsets. There are 2^n possible messages defined on X (in 2^X). In this example, there are 2^4 possible messages. In contrast, fuzzy subsets of X are referred to as *fit vectors* or *fit messages*. Each subset A of X can be defined as one of the continuum-many continuous-valued membership functions μ_A: $X \rightarrow \{0,1\}$. Fuzzy sets can also be represented geometrically and this representation gives more insight into the intricacies of fuzzy sets and operations related to them (Kosko, 2007). According to this representation, the fuzzy power set $F(2^X)$, which is the set of all fuzzy subsets of X, is visualized as a unit hypercube $I^n = [0,1]^n$ and a fuzzy set is any point in the cube I^n. Vertices of the cube I^n define non-fuzzy or crisp sets, which are a subset of X. Thus, crisp sets are special cases of the fuzzy sets.

Fig. 11 shows the geometrical representation of fuzzy sets. The sets on vertices are non-fuzzy sets and long diagonals connect non-fuzzy set complements. A fuzzy set *A* with fit values

$$\left(\frac{1}{3}\frac{4}{5}\right)$$

is represented inside the unit square consisting of all possible fuzzy subsets of two elements. The mid-point of the unit square shown in Fig. 12 is the point of maximum fuzziness. Thus, the proposition for fuzziness can be summarized as follows.

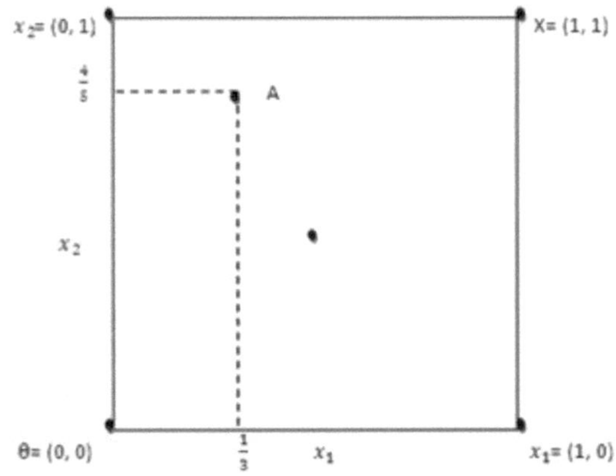

Fig. 11. Geometrical representation of fuzzy set A

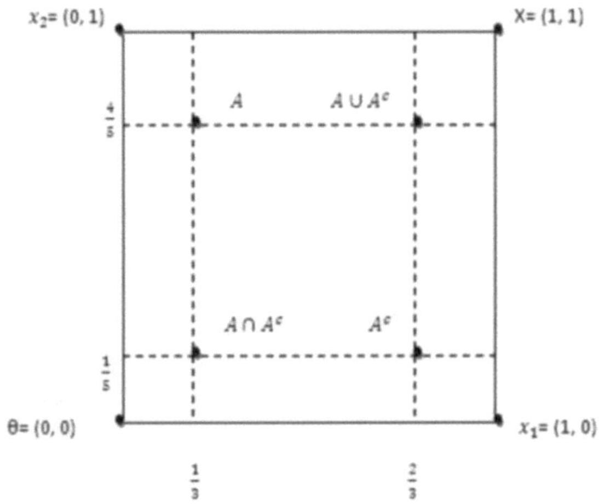

Fig. 12. Fuzzy set A with its complements, under-lap and overlap

A is properly fuzzy if

$$A \cap A^C \neq \varphi \tag{21}$$

$$A \cup A^C \neq X \tag{22}$$

where $A \cap A^C$ and $A \cup A^C$ are termed as overlap and under-lap respectively. The positions of a fuzzy set along with its complement, overlap set and under-lap set are shown in Fig. 12.

With the increase in fuzziness of A, all the 4 points shrink towards the midpoint of the fuzzy square, which is the point of maximum fuzziness. The size or cardinality of a fuzzy set A is given by

$$M(A) = \sum_{i=1}^{n} \mu_A(x_i) \tag{23}$$

Where, $\mu_A(x_i)$ is the membership value of the ith element of n-valued fuzzy set A.

Fuzzy logic has emerged as a promising tool for biological information processing owing to its proximity to human perceptions of logic, and real world situations, which are full of ambiguity. The combination of fuzzy logic and neural networks is reported with promising results in the classification of wines and beverages (Das *et al.*, 1999). The next section presents a novel method of fuzzy pre-processing, which gives simultaneous identification and quantification when the response samples are used to train an ANN classifier. The aim is to tell both the class and the concentration of a sample simultaneously when the sample is presented to an ANN.

5. Fuzzy subsethood for simultaneous identification and quantification of odours

It can be seen from Fig. 4 that the response of the sensor array to almost all the alcohols and alcoholic beverages in the study has saturating tendency at higher concentrations. The quantification task becomes more difficult for mere lack of information at higher concentrations. Both qualitative and quantitative classification tasks coupled together need an integrated approach to be accomplished successfully. As a first step to reduce the complexity of the problem, PCA is used on the raw data.

Fig. 13 shows the PCA plot discriminating different alcohols and alcoholic beverages used in this study. Although PCA can significantly reduce the dimensionality of the data set by having 95% of the variance in first two principal components (PC-1 and PC-2) itself, it has little effect on class separability. This necessitated investigation of a technique, which is based on proper representation of target classes in the output feature space provided the representation is inclusive enough to incorporate in itself both qualitative and quantitative information.

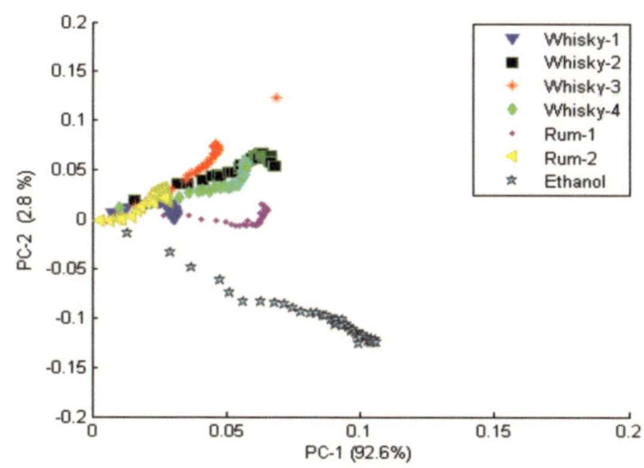

Fig. 13. 2D PCA for sensor array response to 7 alcohols/alcoholic beverages

Fuzzy subsethood representation is such an appropriate technique and applied on the data as follows. The response curves in Fig.4 were sampled at regular intervals of concentration. Each curve was sampled at 48 values of concentration and a total of 336 samples were obtained for 7 gases. As shown in Fig. 14, each sample has two memberships, one to the gas class and the other to the concentration band. For each gas, twelve concentration bands were marked according to increasing no. of drops as band no. 1(b-1) for drops 0-1, band no. 2 (b-2) for drops 1-2 and so on. Each such concentration band consisted of 4 samples.

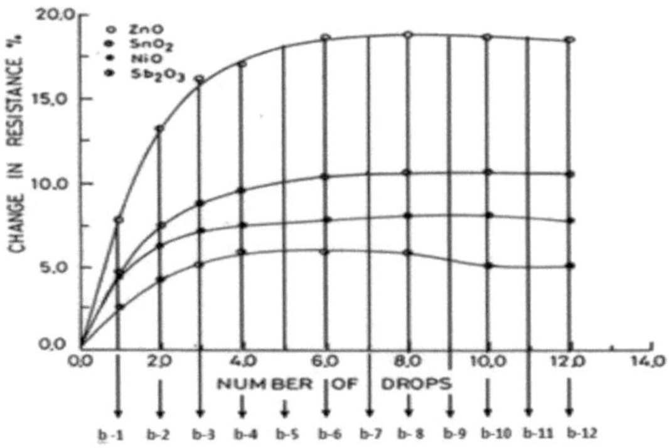

Fig. 14. Concentration bands in sensor response to whisky-2

Fuzzy memberships were assigned to each sensor response sample for all gases as follows. Centroid $\overline{S_j}$ was calculated for each set *j* where *j* denotes a gas and $\overline{S_j}$ has 4 elements as given in

$$\overline{S_j} = \begin{bmatrix} m_{1j} \\ m_{2j} \\ m_{3j} \\ m_{4j} \end{bmatrix} \tag{24}$$

where m_{1j}, m_{2j}, m_{3j} and m_{4j} and represent the centroids for the responses of sensors 1(Sb_2O_3 doped), 2(SnO_2), 3(NiO doped) and 4(ZnO doped) respectively. In this case the centroids are calculated by taking the mean of individual sensor response samples for different alcohols and alcoholic beverages.

The Euclidean distance d_{jk} of the sensor response vector at sample *k* for gas *j* can be obtained in the sensor response ratio vector space as given by

$$d_{jk} = \{(X_{1jk} - m_{1j})^2 + (X_{2jk} - m_{2j})^2 + (X_{3jk} - m_{3j})^2 + (X_{4jk} - m_{4j})^2\}^{\frac{1}{2}} \tag{25}$$

where, x_{1jk} is the sensor response for sensor 1, gas *j*, and sample *k* and so on. Each sample *k* is assigned a membership μ_{jk} in the output feature space, in the fuzzy set A_j for j^{th} gas at sample *k* by using triangular membership function as:

$$\mu_{jnk} = \frac{d_{jnk} - |d_{jnk}|_{\max}}{|d_{jnk}|_{\max} - |d_{jnk}|_{\min}} \tag{26}$$

where, d_{jk} is the Euclidean distance of the sensor response vector at sample *k* for gas *j*. $|d_{jk}|_{\max}$ and $|d_{jk}|_{\min}$ are the modulo of the maximum and minimum values respectively of d_{jk} for a particular gas *j*. Similarly, the sensor response samples obtained from different gases were assigned memberships in different concentration bands of those gases.

Centroid S_{jn} is calculated for each concentration band n of j where j denotes the gas (alcohol and alcoholic beverage type) and S_{jn} has 4 elements as given by

$$\overline{S_{jn}} = \begin{bmatrix} m_{1jn} \\ m_{2jn} \\ m_{3jn} \\ m_{4jn} \end{bmatrix} \tag{27}$$

where, m_{1j}, m_{2j}, m_{3j} and m_{4j} represent the centroids for the responses of sensors 1, 2, 3 and 4 respectively in a concentration band n. The centroids are calculated by taking the simple mean of 4 samples belonging to a particular concentration band. The Euclidean distance d_{jnk} of the sensor response vector at sample k of a gas j and concentration band n of that particular gas can be obtained in the sensor response ratio vector space as given by

$$d_{jnk} = \{(X_{1jnk} - m_{1jn})^2 + (X_{2jnk} - m_{2jn})^2 + (X_{3jnk} - m_{3jn})^2 + (X_{4jnk} - m_{4jn})^2\}^{\frac{1}{2}} \tag{28}$$

where x_{1jk} is the sensor response for sensor 1, gas j, band n, and sample k, and so on. Each sample k of band n is assigned a membership μ_{jnk} in the output feature space, in the fuzzy set A_{jn} again by using triangular membership function as:

$$\mu_{jnk} = \frac{d_{jnk} - |d_{jnk}|_{\max}}{|d_{jnk}|_{\max} - |d_{jnk}|_{\min}} \tag{29}$$

where, d_{jnk} is the Euclidean distance of the sensor response vector at sample k for band n of gas j. $|d_{jnk}|_{\max}$ and $|d_{jnk}|_{\min}$ are the modulo of the maximum and minimum values respectively of d_{jnk} for a particular band n of gas j. It is clear that the fuzzy set A_{jn} is a subset of fuzzy set A_j.

The degree of belongingness of A_{jn} to A_j changes, as n changes for a particular j. Thus, all the elements of A_{jn} can be mapped to a single value, which is the fuzzy subsethood value as defined below, Fuzzy subsethood measures the degree of belongingness of a fuzzy set A to its superset B and is denoted by

$$S(A,B) = Degree(A \subset B) \tag{30}$$

A fuzzy set A can be a subset of another fuzzy set B if $\mu_A(x) \leq \mu_B(x)$ for all x.

The fuzzy-subsethood theorem is given by

$$S(A,B) = \frac{M(A \cap B)}{M(A)} \tag{31}$$

$$S_{jn}(A_{jn}, A_j) = Degree(A_{jn} \subset A_j) \tag{32}$$

Using equations (31) in (32)

$$S_{jn}(A_{jn},A_j)=\frac{M(A_{jn}\cap A_{jn})}{M(A_{jn})} \tag{33}$$

In Fig. 15 the response of the array saturates completely after the concentration band b-4, resulting in a shear lack of information. However, there is a slight change in the response pattern of sensor 1 in band b-9. This change should reflect in the output feature space so that proper quantification can be obtained. For a particular concentration band n consisting of k samples of gas j for the response of sensor i the mean is given as m_{ijn}.

For n = 9, j = 2, and i = 1, the variance V_{ijn} is given by

$$V_{129}=\frac{1}{4\{(X_{1291}-m_{129})^2+(X_{1292}-m_{129})^2+(X_{1293}-m_{129})^2+(X_{1294}-m_{129})^2\}} \tag{34}$$

From (28) the Euclidean distance d_{jnk} of sample k = 1 of j = 2, n = 9, is given by

$$d_{291}=\{(X_{1291}-m_{129})^2+(X_{2291}-m_{229})^2+(X_{3291}-m_{329})^2+(X_{4291}-m_{429})^2\}^{\frac{1}{2}} \tag{35}$$

It can be observed that the first term of the variance V_{129} finds itself as a component of the Euclidean distance d_{jnk}. Since the variance V_{ijn} is calculated for the response samples of a single sensor i and the Euclidean distance d_{jnk} of a particular sample takes into account the responses of all the 4 sensors of the array, any significant change in the response of any of the 4 sensors in a concentration band is certainly going to reflect in the Euclidean distance of any sample for the same concentration band. Since membership values and subsethood are primarily based upon Euclidean distance, the change in variance for a particular sensor in a concentration band will play a part in the subsethood calculation for the response vector obtained from all the 4 sensors in the same concentration band. In this way, if any one of the 4 sensors shows less saturation at higher concentrations the possibility of correct quantification increases.

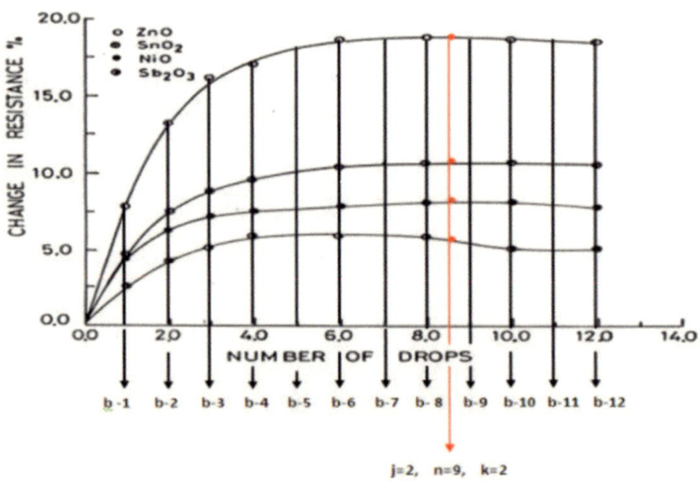

Fig. 15. Second sample of ninth band of whisky-2

5.1 Simulation results and discussions

The subsethood values were obtained for each concentration band of a particular gas and were used as the target for the neural network classifier. For a total of 7 alcoholic beverages 7 neurons were kept in the output layer. A neuron corresponding to a particular gas class was supposed to fire at a value corresponding to the fuzzy subsethood of the particular concentration band, to which the test sample belonged while all other neurons were supposed to be deactivated. A tolerance of 2% for the target fuzzy subsethood was considered appropriate. A single hidden layer feed-forward ANN was trained with a BP algorithm. The input layer consisted of 4 neurons and the number of neurons in the hidden layer was optimized by experimentation. The input data were divided into training and testing data matrices. The simulations were carried out on MATLAB platform and several different versions of a BP algorithm available in the MATLAB neural network toolbox (Mathswork Inc., 2007) were tested. Three training methods based upon a BP algorithm namely *Trainoss, Trainscg* and *Trainlm* have been found to give satisfactory results. To eliminate the possibility of over fitting m-fold cross validation scheme (Hykin, 2009) was used. For all the three versions *logsigmoidal* activation function was used. All the three training methodologies use default values of learning rate η and momentum constant α adaptively during the simulation run. The number of neurons in the hidden layer of the network was varied from two to nine and system error (mean square error) was noted. The networks were trained to a fixed 10,000 epochs with an error goal of 0.0001. *Trainoss, Trainscg* and *Trainlm* are found to train the network best when the number of neurons in the hidden layer of the network was 7, 5, and 6 respectively. In the testing phase, 12 samples were taken for a particular beverage with one sample each from a particular concentration band. The proposed network was found to give the best testing phase performance when the network was trained with *Trainlm* methodology. Table 1 shows the summary of best classifications achieved qualitatively and Table 2 shows the quantitative classification results for a network with an optimized topology of 4:6:7 trained with *Trainlm*. For qualitative classification 83 out of 84 samples were identified correctly giving a result of 98.97%.

Gas class	No. of samples correctly detected out of 12 for each gas class
Whisky-1	12
Whisky-2	12
Whisky-3	12
Whisky-4	11
Rum-1	12
Rum-2	12
Ethanol	12
Total % classification achieved	98.97

Table 1. Results of qualitative classification

Gas class	No. of samples correctly detected in Concentration Bands b-1 to b-12
Whisky-1	8
Whisky-2	8
Whisky-3	6
Whisky-4	9
Rum-1	6
Rum-2	9
Ethanol	10
Total % quantification achieved	66.67

Table 2. Results of quantitative classification

Whereas, 56 out of 84 samples were detected correctly in concentration bands b-1 to b-12, giving a success rate of 66.67%. The results seem to be very encouraging since the sensor response at higher concentrations of the test gas remains saturated for almost all types of alcohols and alcoholic beverages, resulting in a shear lack of information at higher concentrations as evident from Fig. 4.

6. Conclusions

In this chapter a neural fuzzy identifier/quantifier was presented for discrimination of several alcoholic beverages using responses of a poorly selective sensor array. The simulation results obtained using fuzzy subsethood based feature extraction, validate the presumption that the limitations imposed by poor selectivity of chemical sensors can be overcome using appropriate soft computational technique. This chapter also highlights the importance of a pre-processing stage before the response sampled are fed to a neural classifier. The proposed technique of fuzzy subsethood encoding is also similar to a pre-processing stage, which makes the subsequent neural classification faster and error free. It is important to have a classifier with a small number of neurons in the hidden layer so that it can be implemented easily into custom VLSI chips. The technique presented in this chapter accomplishes the identification/quantification task with a few neurons in the hidden layer (i.e. 6) and hence its efficacy is established.

There is a scope for future work by trying to make the identification/quantification techniques less problem-dependent and more general in nature, which would eventually enable the realization of a highly marketable hand-held E-nose system.

7. Acknowledgements

The author is deeply indebted to Prof. R. Dwivedi and Prof. R. R. Das of the Centre for Research in Microelectronics (CRME), IIT-BHU, India for providing useful data and sensor fabrication facilities. Thanks are due to IEEE Sensors Journal for permitting the reproduction of a couple of Figures from previous publications of the author. The author is thankful to Mr. Somesh Kumar for his help in preparation of this manuscript.

8. References

Bishop, C.M. (1994). *Neural Networks for Statistical Pattern Recognition,* Oxford University Press, Oxford, UK

Chaturvedi, A.; Mishra, V.N.; Dwivedi, R. and Srivastava, S.K. (1999). Response of oxygen plasma treated thick film tin oxide sensor array for LPG, CCl_4, CO and C_3H_7OH, Microelectronics Journal, Vol. 30, pp. 259-264

Cover, T.M. (1965). Geometrical and Statistical Properties of Systems of Linear Inequalities with Applications In Pattern Recognition, IEEE Trans. Electron. Comput., EC-14, 326-334

Das, R.R.; Shukla, K.K.; Dwivedi, R. and Srivastava, A.R. (1999). *Discrimination of Individual Gas/Odor Using Responses of Integrated Thick Film Tin Oxide Sensor Array and Fuzzy-Neuro Concept,* Microelectronics Journal, Vol. 30, pp. 793-800

Gardner, J.W. (1987). *Pattern Recognition in the Warwick Electronic Nose,* 8th International Congress of European Chemoreception Research Organization, University of Warwick, UK

Hykin, S. (2009). *Neural Networks and Learning Machines,* Chapter 5, PHI

Ikohura, K. (1981). *Tin Oxide Gas Sensor for Deoxidizing Gas, New Materials and New Processes in Electrochemical Technology,* Vol. 1, pp. 43-50

Kosko, B. (2007). Neural Networks and Fuzzy Systems: A Dynamical Systems Approach to Machine Intelligence, India: Prentice Hall, Chapter 7

Kumar, R.; Das, R.R.; Mishra, V.N. and Dwivedi, R. (2010). A neuro-fuzzy classifier-cum-quantifier for analysis of alcohols and alcoholic beverages using responses of thick film tin oxide gas sensor array, IEEE Sensors Journal, Vol. 10, No. 9, pp. 1461-1468

MATLAB (2007). *Toolbox Documentations,* MATHWORKS Inc., USA, Licence Number: 338828-DVDWIN07B

Menini, A.; Lagostena, L. and Boccaccio, A. (2004).*Olfaction: From Odorant Molecule to the Olfactory Cortex,* New Physical Sci., Chapter 19, pp. 101-104

Morrison, S.R. (1987). *Selectivity in Semiconductor Gas Sensors,* Sensors and Actuators, Vol. 12, pp. 425 440

Moseley, P.T. (1992). *Material Selection for Semiconductor Gas Sensors,* Sensors and Actuators B, Vol. 6, pp. 149-156

Nagle, H. T.; Schiffman, S. S. and Osuna, R.G. (1998) *The How and Why of Electronic Noses,* IEEE Spectrum, pp. 22-34

Nakamoto, T.; Fukunishi, K. and Morizumi, T. (1990). *Identification Capability of Odor Sensor Using Quartz Resonator Sensor Array and Neural-Network Pattern Recognition,* Sensors and Actuators B, Vol. 1, pp. 473-476

Nayak, M.S.; Dwivedi, R. and Srivastava, S.K. (1992). Integrated Sensor Array for Detection of Alcohols and Alcoholic Beverages, J. Inst. Eng. (India), Vol. 75, pp. 94-98

Osuna, R. G. (2002).*Pattern Analysis for Machine Olfaction: A Review,* IEEE Sensors Journal, Vol. 2, pp. 189-202

Sarro, P.M. (1992). *Sensor Technology Strategy in Silicon,* Sensors and Actuators A, Vol. 31, pp. 138-143

Sears, W.M.; Colbow, K.; Slamka, R. and Consadori, F. (1990). *Selective Thermally Cycled Gas Sensing Using Fast Fourier Transform Technique,* Sensors and Actuators B, Vol. 2, pp. 283-289

Shurmer, H.V.; Gardner, J.W. and Corcoran, P. (1990). *Intelligent Vapor Discrimination Using a Composite Twelve Element Sensor Array,* Sensors and Actuators B, Vol. 1, pp. 256-260

Watson, J. (1984). *The Tin Oxide Gas Sensors and Its Applications,* Sensors and Actuators, Vol. 5, pp. 29-42

Zadeh, L.A. (1965). Fuzzy sets, Inf. Control, Vol. 8, pp. 338-353